The Cation–π Interaction

Shinji Yamada

The Cation–π Interaction

Springer

Shinji Yamada
Department of Chemistry
Ochanomizu University
Tokyo, Japan

ISBN 978-981-19-7334-5 ISBN 978-981-19-7335-2 (eBook)
https://doi.org/10.1007/978-981-19-7335-2

This Springer imprint is published by the registered company Springer Nature Singapore Pte Ltd.
The registered company address is: 152 Beach Road, #21-01/04 Gateway East, Singapore 189721, Singapore

Preface

A cation-π interaction is a non-covalent bond between a cation and a π-plane, which contributes to the construction of two- and three-dimensional structures of biomolecules such as proteins, enzymes, and nucleic acids, as well as various organic compounds, polymers, and composite materials. Such higher-order structures are responsible for their properties and functions. In addition, cation-π interactions play an important role in both biological processes and chemical reactions.

The term "cation-π interaction" was coined about 30 years ago. Despite this short history in comparison to the study of other non-covalent interactions, the properties and roles of cation-π interactions have attracted much attention, and significant progress has been made in our understanding of their roles and applications. In particular, the recent developments in the field of materials science have been remarkable. The importance of cation-π interactions across such a wide range of fields is due to their large interaction energies and the diversity of the systems in which they are observed.

There are many different types of non-covalent interactions, and it is rarely the case that only one interaction is involved in a system. In complex systems, in particular, it is not difficult to imagine that multiple interactions are involved in a complex manner. It is also known that some of these interactions work in concert. This book will focus on examples in which a cation-π interaction is considered to be the primary interaction.

Chapter 1 outlines the position of cation-π interactions within the framework of non-covalent interactions as a whole, while Chap. 2 describes the characteristics and features of cation-π interactions and Chap. 3 provides their role in biochemical processes found in nature. Chapter 4 presents examples of synthetic reactions based on control of the conformation and arrangement of organic molecules through cation-π interactions in the liquid and solid phases, respectively. Chapter 5 introduces the properties and functions of polymers and composite materials involving cation-π interactions.

The fields of biochemistry, organic chemistry, polymer chemistry, and materials science are borderless, especially when viewed from the perspective of interactions. For example, the interaction between quaternary ammonium ions and benzene rings plays key roles in acetylcholine recognition, mussel underwater adhesion, polymer adhesion, stereoselectivity of organic reactions, and recognition-based functional materials. The author will be delighted if this book can contribute not only to the further development of each field but also to the development of new scientific fields by providing an overview of the wide range of scientific fields covered by the keyword "cation-π interaction".

I am deeply indebted to S. Tsuzuki (AIST) for his many valuable comments on Chap. 2 of the manuscript. Moreover, my interest in the role of cation-π interactions in organic reactions was first inspired by discussion with him. It can be said, therefore, that this book might never have been born without my meeting Dr. Tsuzuki.

Tokyo, Japan Shinji Yamada

Contents

Chapter 1
Introduction

1.1 Non-covalent Interaction

Covalent bonds are chemical bonds in which the atoms bind to each other to form molecules, whereas non-covalent bonds are bonds that do not involve the sharing of electrons, with the non-covalent interactions between molecules leading to the formation of complexes. As such, they are weaker than covalent bonds and are characterized by the reversibility of complex formation and cleavage of the complexes to their constituent molecules (Fig. 1.1). As the electronic properties of the two molecules affect the nature of the interaction, various types of non-covalent interactions with unique structural features are known to occur.

Such intra- and intermolecular interactions lead to the formation of thermodynamically stable conformers, supramolecules, host–guest complexes, and self-assembled nanostructures. For example, the primary structure of enzymes and proteins is determined by covalent bonds, while the secondary, tertiary, and even quaternary structures are determined by hydrogen bonds [1] and other non-covalent interactions. These higher-order structures play a variety of important roles in proteins and enzymes.

Non-covalent interactions can be classified into two categories: One is based on the structural features of the interaction systems, and the other is based on the physicochemically defined forces [2]. The hydrogen bond, halogen bond, π–π, CH–π, cation–π, and anion–π interactions are classified as structure-based interactions (Fig. 1.2), which are useful for our understanding of the role of the interaction on the geometries of the respective molecules and complexes. On the other hand, the electrostatic, inductive, dispersion, charge-transfer, and repulsive forces are physicochemical forces, which are components of the structure-based interactions, and are important to our understanding of the nature of the interactions. For example, in the case of hydrogen bond, the directional dependence and the large interaction energy are mainly due to the contribution of electrostatic force [3].

S. Yamada, *The Cation-π Interaction*,
https://doi.org/10.1007/978-981-19-7335-2_1

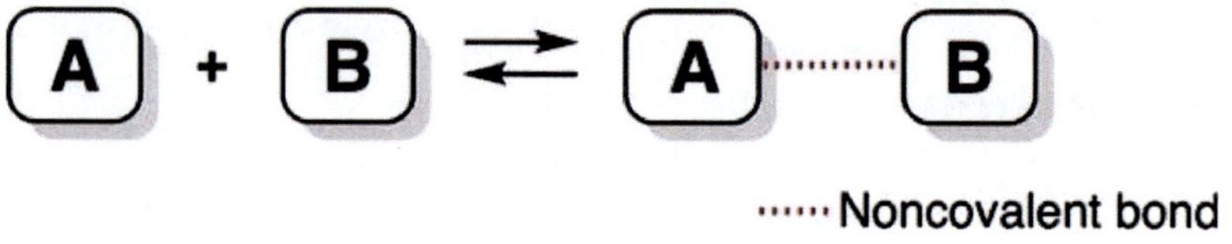

Fig. 1.1 Schematic representation of a non-covalent bond (A and B represent organic molecules)

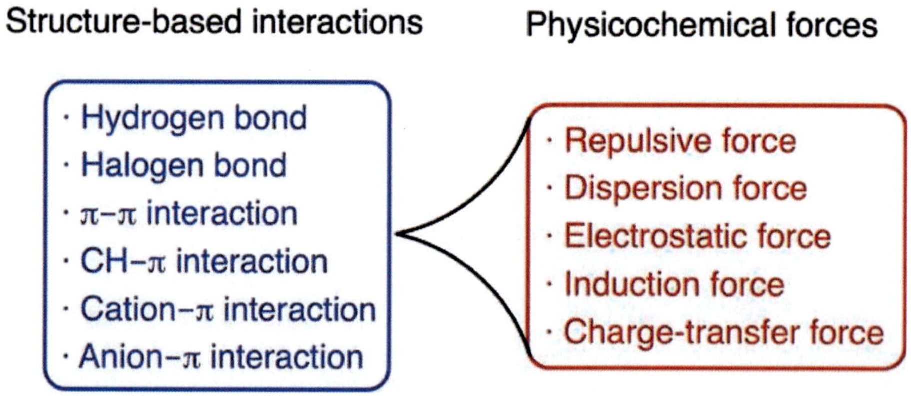

Fig. 1.2 Structure-based interactions, and physicochemical forces, which are components of structure-based interactions

1.2 Cation–π Interaction

Among the various structure-based interactions, the cation–π interaction [4] has received considerable attention over the last two decades. Cation–π interactions were first identified in 1981 by Kabarle et al. [5] who elucidated the existence of attractive non-covalent interactions between alkali metal ions and aromatic compounds in the gas phase. Their interaction energies were reported to be about −19 ~ −38 kcal/mol. In 1985, Meot-Ner reported that quaternary ammonium ions also interact effectively with the π-systems[6]. These non-covalent interactions between a cation and a π-plane were initially referred to as "polar–π" or "ion–dipole" [7]. Dougherty subsequently proposed a more general and understandable term, "cation–π interaction," in 1990,[8] and today this term has become widely accepted.

The origin of the attraction in the cation–π interaction has been shown to be electrostatic force. Figure 1.3 shows the electrostatic potential map of a benzene–ammonium complex. The carbon atoms of the benzene ring are slightly negative, while the hydrogen atoms are slightly positive; therefore, there is a negative charge in the center of the benzene. The electrostatic force acting between this negative charge and the positive charge of the cation stabilizes the structure of the cation in contact with the central part of the benzene [7]

There are several interaction systems involving a benzene ring, such as π–π, CH–π, and cation–π interactions. The major force in π–π [2] and CH–π [9] interactions

Fig. 1.3 Electrostatic potential map for a benzene–ammonium complex. Reprinted with permission from [7]. Copyright 2013 American Chemical Society

is dispersion force, whereas those in cation–π interactions are electrostatic [4, 7] and inductive forces [10, 11] due to the presence of the cation. These differences in the dominant forces explain the larger interaction energy of cation–π interactions than those of π–π and CH–π interactions. Thus, by knowing which force is the major contributor in each interaction, we can understand the nature of the structure-based interaction.

Cation–π interactions have attracted much attention in the field of biological chemistry, including structural biology and neurobiology, as well as in the biochemical processes in various organisms. In particular, cation–π interactions are often involved in the three-dimensional protein structures and in the recognition of biological substrates [12]. This interaction plays an important role not only in biological systems, but also in artificial systems such as synthetic chemistry [13] and materials science [14] due to the following advantages.

(1) A variety of cationic species including cationic intermediates, transition states as well as excited states can interact with π-systems.
(2) The π-systems contain a variety of π-components such as aromatic rings, alkenes, alkynes, carbonyls, and conjugated systems.
(3) The interaction energy is generally larger than that of other π-interactions.

The diversity of cations and π–components in the cation–π interaction systems (Fig. 1.4), and the larger interaction energy make cation–π interactions a promising

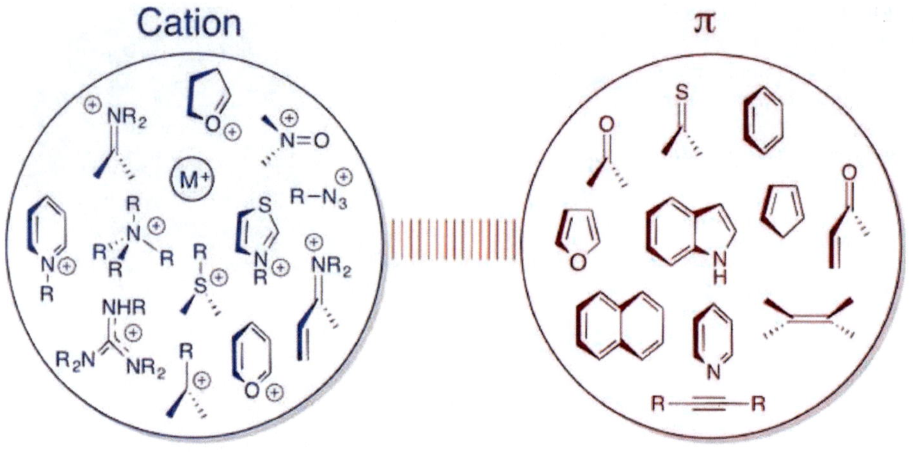

Fig. 1.4 Cation–π interactions between various cations and π-systems. Reprinted with permission from [13]. Copyright 2018 American Chemical Society

tool for the development of various fields of science in both the liquid and solid phases [15], as well as in solution.

In this book, the author will introduce the cation–π interaction from its fundamental nature to its role in biological systems and applications in synthetic chemistry and materials science, which will provide insights into the discovery of new phenomena and the development of new fields of science.

References

1. Pauling L, Corey RB, Branson HR (1951) The structure of proteins: two hydrogen-bonded helical configurations of the polypeptide chain. Proc Natl Acad Sci USA 37:205–211
2. Hunter CA, Sanders JKM (1990) The nature of π-π interactions. J Am Chem Soc 112:5525–5534
3. Gordon MS, Jensen JH (1996) Understanding the hydrogen bond using quantum chemistry. Acc Chem Res 29:536–543
4. Ma JC, Dougherty DA (1997) The cation−π interaction. Chem Rev 97:1303–1324
5. Sunner J, Nishizawa K, Kebarle P (1981) Ion-solvent molecule interactions in the gas phase: the potassium ion and benzene. J Phys Chem 85:1814–1820
6. Deakyne CA, Meot-Ner (Mautner) M (1985) Unconventional ionic hydrogen bonds: 2. $NH^{+}\cdots\pi$ complexes of onium ions with olefins and benzene derivatives. J Am Chem Soc 107:474–479
7. Dougherty DA (2013) The cation-π interaction. Acc Chem Res 46:885–893
8. Stauffer DA, Barrans RE Jr, Dougherty DA (1990) Biomimetic catalysis of an S_N2 reaction resulting from a novel form of transition-state stabilization. Angew Chem Int Ed Engl 29:915–918
9. Tsuzuki S, Fujii A (2008) Nature and physical origin of CH/π interaction: significant difference from conventional hydrogen bonds. Phys Chem Chem Phys 10:2584–2594
10. Tsuzuki S, Yoshida M, Uchimaru T, Mikami M (2001) The origin of the cation/π interaction: the significant importance of the induction in Li^{+} and Na^{+} complexes. J Phys Chem A 105:769–773
11. Tsuzuki S, Mikami M, Yamada S (2007) Origin of attraction, magnitude, and directionality of interactions in benzene complexes with pyridinium cations. J Am Chem Soc 129:8656–8662

12. Mahadevi AS, Sastry GN (2013) Cation-π interaction: its role and relevance in chemistry, biology, and material science. Chem Rev 113:2100–2138
13. Yamada S (2018) Cation-π interactions in organic synthesis. Chem Rev 118:11353–11432
14. Zhao G, Zhu H (2020) Cation-π interactions in Graphene-containing systems for water treatment and beyond. Adv Mater:1905756
15. Yamada S (2020) Cation–π interactions in organic crystals. Coord Chem Rev 415, 213301. Article No S0010854519307702

Chapter 2
Fundamentals of Cation–π Interactions

2.1 Metal Ions

2.1.1 Alkali Metal Ions

Figure 2.1a shows a schematic representation of the cation–π interaction between a metal ion and a benzene ring, in which a generic metal cation is positioned over the benzene ring along the sixfold axis with the distance of R. The calculated optimized geometry of a $K^+\cdots$ benzene complex is shown by the space-filling model in Fig. 2.1b, which demonstrates that K^+ is on the center of the benzene ring with the van der Waals contact [1].

The interaction energies between alkali metal ions and benzene in the gas phase have been determined experimentally. The interaction energies of Li^+, Na^+, and K^+ with benzene are −38.3, −28.0, and −19.2 kcal/mol, respectively [2]. Theoretical calculations have also been performed and are in good agreement with experimental results (Table 2.1). A comparison of monovalent cations reveals that as the ionic radius increases, the cation-benzene distance increases and the interaction force decreases. Smaller cations can get closer to the ring, thus producing larger electrostatic energy and inducing larger polarization in the benzene ring, which results in a larger interaction energy [3, 4].

Figure 2.2 shows the relationship between the distance between a lithium ion and a benzene ring and the interaction energy. It is clear that the attractive force acts from a significant distance, and the energy (E_{MP2}) becomes most stable at 1.87 Å. The electrostatic force (E_{es}) and the induction force (E_{ind}) increase as the distance decreases, indicating that they both make a significant contribution to E_{MP2} [3]. The contributions of the charge-transfer force appear to be small as this interaction is effective only at short distances. As shown in Table 2.1, the electrostatic force (E_{es}) and the induction force (E_{ind}) contribute significantly to stabilization in all cases, indicating that the cation–π interactions are characterized by electrostatic and induction forces.

S. Yamada, *The Cation-π Interaction*,
https://doi.org/10.1007/978-981-19-7335-2_2

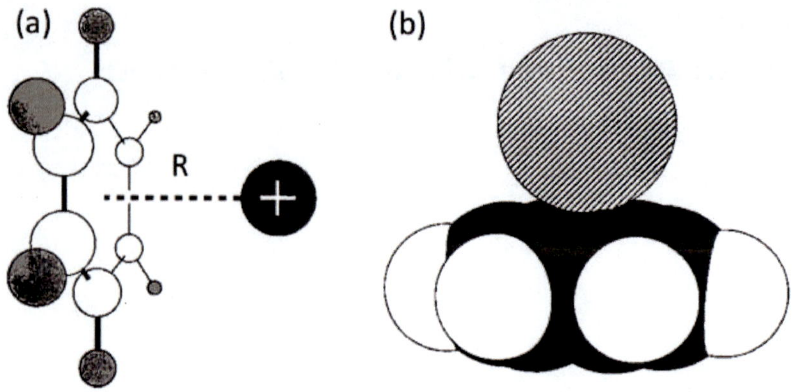

Fig. 2.1 Schematic representations of a cation–π interaction: **a** the interaction showing a generic cation positioned over a benzene ring along the sixfold axis and **b** a space-filling model of the $K^+\cdots$ benzene complex at its optimized geometry. Reprinted with permission from [1]. Copyright 1997 American Chemical Society

Table 2.1 Interaction energies between metal cations and benzene

Cluster	E_{total}[a]	E_{es}	E_{ind}	R/Å[b]
Li^+/C_6H_6	−35.4	−18.7	−46.9	1.869
Na^+/C_6H_6	−21.3	−14.8	−21.0	2.425
K^+/C_6H_6	−17.0	−11.9	−12.8	2.805
Rb^+/C_6H_6	−13.9	−9.2	−8.4	3.165
Cs^+/C_6H_6	−12.1	−7.9	−6.4	3.414
Mg^{2+}/C_6H_6	−109.9	−37.8	−162.4	1.958
Ca^{2+}/C_6H_6	−73.7	−31.2	−91.2	2.361

[a] All energies are shown in kcal/mol. [b] R indicates the distance between the centroid of the benzene and the center of the metal ion

Calculations performed in consideration of water have revealed that the interaction energy decreases as the coordination number of the water molecules increases: The interaction energy decreased by −37.77, −25.22, −18.07, and −9.94 kcal/mol as the number of coordinating water molecules in the $Li^+(H_2O)_n\cdots$ benzene complex increased from 0 to 3 [5]. This may be due to the weakening of E_{es} and E_{ind} by the coordination of water molecules; therefore, it should be noted that the interaction energies in an aqueous solution differ markedly from those in the gas phase.

The formation of alkali metal ion-π complexes has been elucidated by X-ray structural analyses. Gokel et al. [6] synthesized a lariat ether, in which an aromatic ring is attached to an azacrown ether via a spacer, and investigated its ability to uptake potassium ions (Fig. 2.3a). Compound **1a** bearing a 4-hydroxybenzene group and KI form a complex **2a**. The X-ray crystal structure of **2a** shows that the potassium ion is incorporated into the crown ether and the aromatic rings are covered with a lid on the top and bottom through cation–π interactions, while the iodide ion is

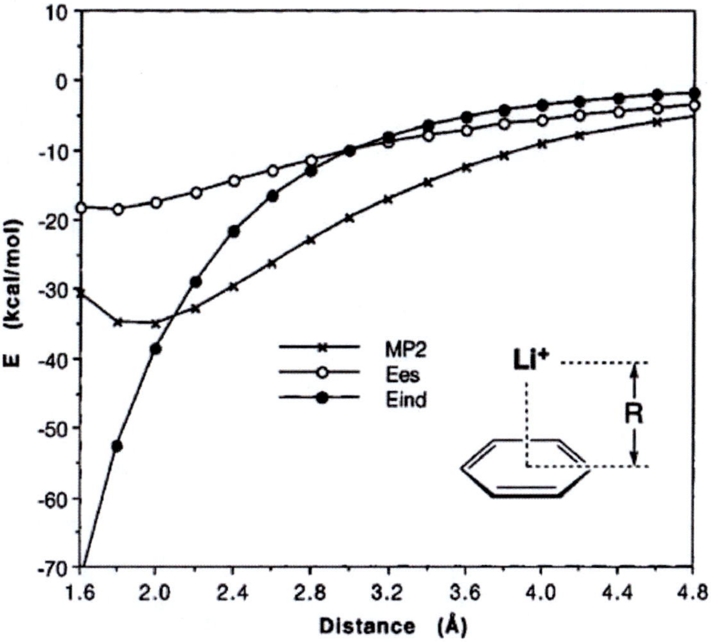

Fig. 2.2 MP2 interaction energies and electrostatic and induction energies (E_{es} and E_{ind}) in a benzene$\cdots$$Li^+$ complex. Reprinted with permission from [3]. Copyright 2001 American Chemical Society

located outside the ring (Fig. 2.3b). The distance between the potassium ion and the benzene ring is 3.43 Å. On the other hand, complex **2b** with a pentafluorophenyl group has an open lid structure and the iodide ion is located on the potassium ion in the azacrown ring instead of on the benzene ring (Fig. 2.3c). This is probably due to the electrostatic repulsion between the pentafluorophenyl group and the potassium ion. Other azacrown ethers possessing vinyl [7] and propargyl [8] moieties also form similar complexes, in which the side arms interact with the potassium ion through cation–π interactions.

2.1.2 Alkaline Earth Metal Ions

The interaction forces of alkaline earth metal ions, Mg^{2+} and Ca^{2+}, with benzene are much stronger than those of alkali metal ions: The interaction forces of Mg^{2+} and Ca^{2+} with benzene are −109.9 and −73.7 kcal/mol, respectively [4]. As can be seen from Table 2.1, these interactions are characterized by the fact that the induction forces are far larger than the electrostatic forces, and the E_{ind} for Mg^{2+} is more than four times larger than the E_{es}. On the other hand, the value decreases significantly with the coordination of water molecules similar to the case of Li^+ described above [5].

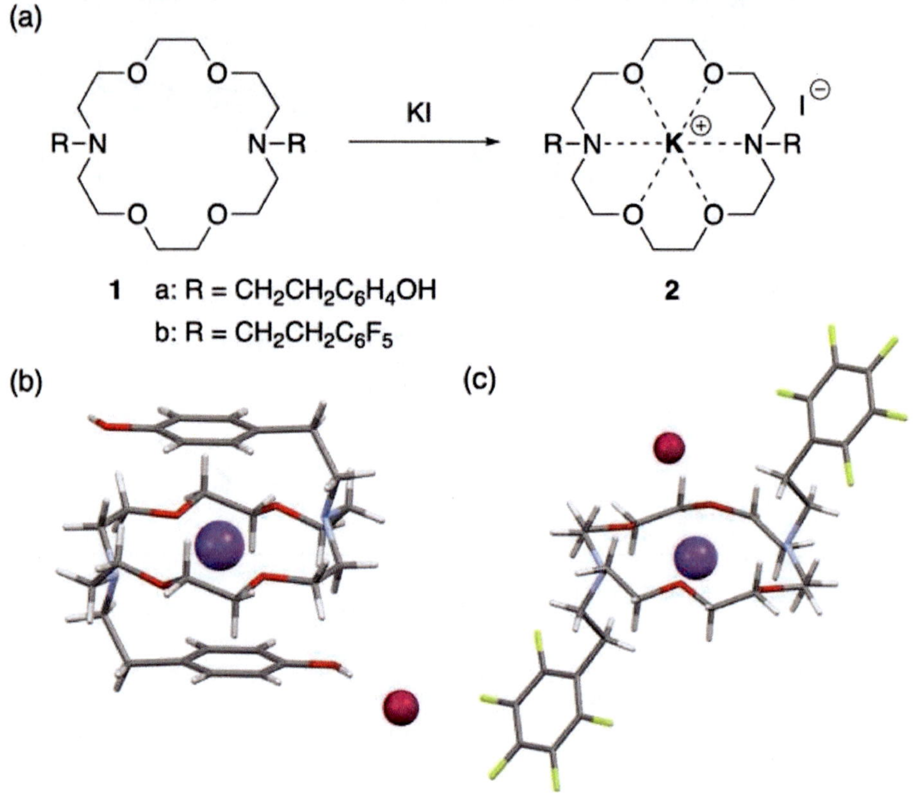

Fig. 2.3 **a** Formation of complex **2** from azacrown ether **1** with KI, and the crystal structures of **b** **2a** and **c** **2b**

2.1.3 Transition Metal Ions

It is known that there is a large interaction force between transition metal ions and aromatic rings. For example, the interaction energy between Cu^+ and benzene is −48.7 kcal/mol, which is comparable to that for the lithium ion. The geometry in this interaction is similar to a cation–π interaction, but it has been shown that the main force of attraction in this case is the interaction between the d orbital of Cu^+ and the π* orbital of benzene [9]. Such orbital interactions are very different from the characteristic features of cation–π interactions, in which E_{es} and E_{ind} are dominant, and are generally not included among cation–π interactions. Therefore, a quantitative evaluation is necessary to clarify whether the contribution of the orbital-to-orbital interaction or E_{es} and E_{ind} is important.

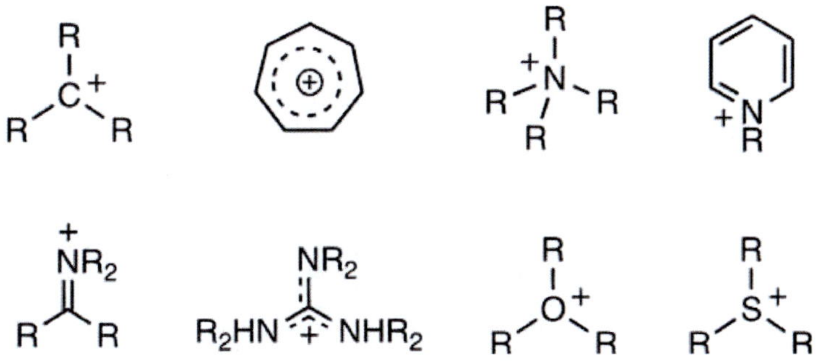

Fig. 2.4 Various organic cations

2.2 Organic Cations

2.2.1 Structural Diversity

A variety of organic cations having a positive charge on the C, N, O, or S atom interact with π-surfaces (Fig. 2.4). While carbocations together with tropylium, pyridinium, guanidinium, iminium, oxonium, and sulfonium ions have a planar structure, quaternary ammonium ions have a three-dimensional structure (Fig. 1.3). The interaction energies of such organic cations with a π-plane are generally smaller than those of alkali metal cation–π interactions. However, organic cations have an advantage in terms of diversity as cations such as onium ions, cationic intermediates, and cationic transition states are effective for these interactions. Such structural diversity greatly expands their use in various fields. The structure of the substrates significantly affects the interaction energies, but various experimental and theoretical calculations have reported that the values are in the range of a few to several dozen kcal/mol (see, Sects. 2.2.2 and 2.2.3), which are much larger than those other interactions involving π-electrons. Similar to alkali metal ions, organic cation–π interactions have also been shown to be dominated by electrostatic and induction forces. Among the various organic cations, the typical cations pyridinium, quaternary ammonium, and guanidinium ions are discussed and introduced below.

2.2.2 Pyridinium Ions

Dougherty et al. [10] showed that cyclophane **3a** exerts a larger stabilizing energy in the incorporation of 1- and 2-methylquinolinium salts than in that of the corresponding quinolines (Fig. 2.5). Furthermore, **3b** preferentially incorporates cations in aqueous solution, overcoming solvation by water molecules. This suggests that the uptake of quinolinium is due to a cation–π interaction rather than a π–π interaction.

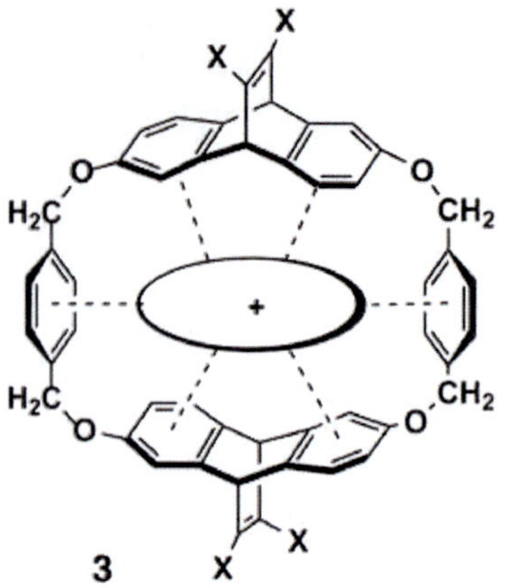

3

a: X = CO_2CH_3
b: X = CO_2^- Cs^+

Guest	-ΔG°(**3a**)	-ΔG°(**3b**)
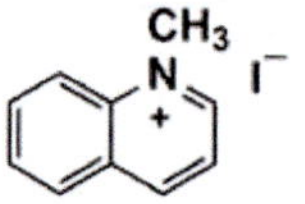	3.5	7.6
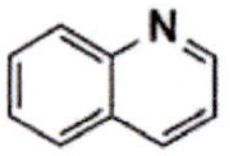	0.0	5.4
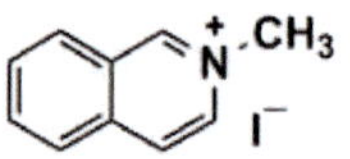	2.5	7.2
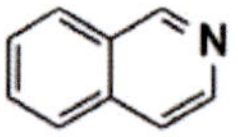	0.2	6.3

[a] kcal/mol

Fig. 2.5 Stabilization energies for host–guest complexes

Figure 2.6 shows the energy diagram of benzene with *N*-methylpyridinium. It can be seen that the attractive force is acting from a marked distance. In this complex, the face-to-face form is the most stable with a value of −9.36 kcal/mol at a distance of 3.6 Å [11].

Face-to-face interaction is often confused with a charge-transfer interaction due to its geometrical features, but the energy diagram shows that the main force is not the orbital interaction as the interaction is observed from a marked distance. It should be noted that the similarity of the geometrical structure to that of the π–π interaction does not necessarily correspond to any similarity in their natures.

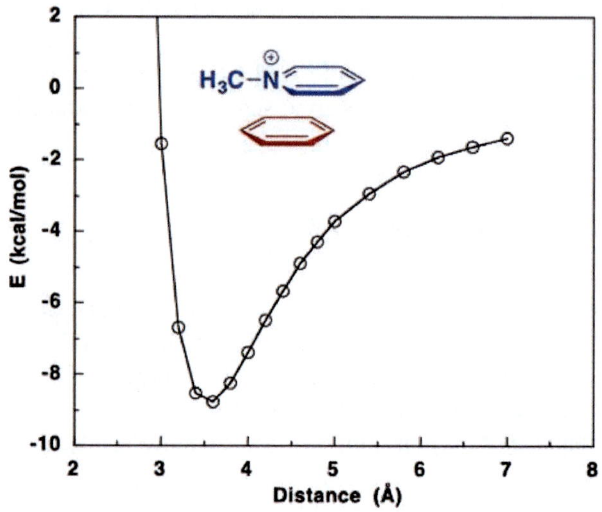

Fig. 2.6 MP2/cc-pVTZ level interaction energy potential for a N-methylpyridinium-benzene complex. Reprinted with permission from [11]. Copyright 2007 American Chemical Society

Table 2.2 Interaction energies for benzene complexes[a, b]

Complex	E_{total}	E_{es}	E_{ind}	E_{rep}	E_{corr}
C_6H_6/C_6H_6	−2.48	0.90	−0.25	3.01	−6.14
C_6H_6/C_5H_5N	−3.04	0.39	−0.22	3.27	−6.48
$C_6H_6/C_5H_5NH^+$	−14.77	−8.12	−9.13	8.12	−5.63
$C_6H_6/C_5H_5NMe^+$	−9.36	−4.05	−3.52	5.79	−7.58
C_6H_6/K^+	−17.20	−11.90	−12.80	11.80	−4.40

[a] MP2/6-311G*, [b] In kcal/mol

Table 2.2 shows the interaction energies (E_{total}) of benzene with benzene, pyridine, pyridinium, *N*-methylpyridinium, and potassium ions, as well as the interaction energies determined by energy decomposition analysis [11]. It is clear from Table 2.2 that the interaction energies (E_{total}) of the pyridinium and *N*-methylpyridinium ions with benzene are much larger than the interaction energies of benzene-benzene and benzene-pyridine (π-π interactions). The interaction between pyridinium and benzene is dominated by electrostatic (E_{es}) and induction (E_{ind}) interactions, as observed in the $K^+\cdots$benzene complex, which is consistent with the characteristics of cation–π interactions. It is important to note that other attractive forces are not entirely uninvolved. In comparison with the $K^+\cdots$benzene complex, the pyridinium$\cdots$benzene complex is characterized by the inclusion of larger dispersion forces as well.

In the solid and liquid states, charge-transfer absorption bands are often observed between pyridinium and electron-rich compounds. However, the observation of a charge-transfer absorption band does not necessarily mean that the charge-transfer interaction is dominant in the system. It has been reported that charge-transfer interactions are rarely involved in intermolecular association or coordination control in solution [12–14].

An example of the interaction between *N*-methylpyridinium and a benzene ring is reported by Yamada et al. [15]. Figure 2.7 shows the X-ray crystal structures of nicotinic amide **4** and the corresponding *N*-methyl salt **5**. Comparing these crystal structures, it is clear that the two rings overlap in a face-to-face manner in the *N*-methyl salt **5**, whereas the two rings are separated from each other in **4**. This geometrical difference strongly suggests the existence of a cation–π interaction in **5**. This interaction is used for face-selective nucleophilic addition reaction to a pyridinium ring (see, Chap. 4). It should be noted that not only intramolecular interactions but also intermolecular cation–π interactions with neighboring molecules were observed in the crystal structure, leading the formation of an infinite columnar motif (see, Sect. 2.5.1.5).

Hunter et al. [16, 17] determined the interaction energies by using the chemical double-mutant method with four complexes **A–D**, consisting of four compounds **6–9** (Fig. 2.8). The contribution of the cation–π interaction in **A(6**$\cdots$**7)** was evaluated by comparison of the association constants of **A(6**$\cdots$**7)** with those of other

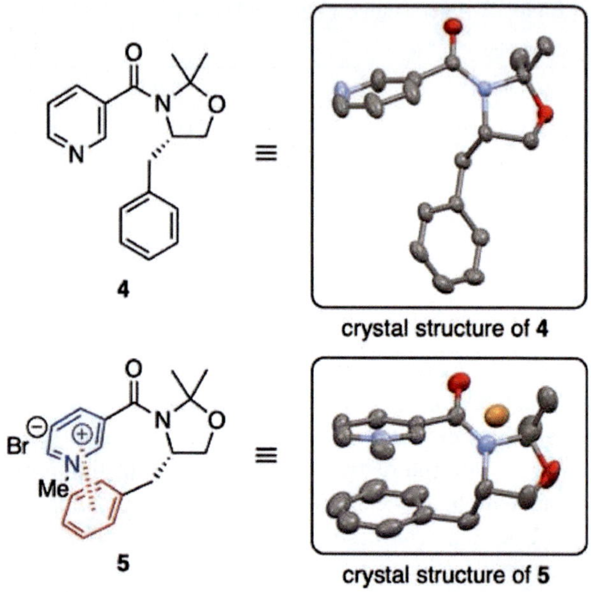

Fig. 2.7 Crystal structures for nicotinic amide **4** and the corresponding *N*-methylpyridinium salt **5**. Reprinted with permission from [50]. Copyright 2018 American Chemical Society

complexes **B(8**···**7)**, **C(6**···**9)**, and **D(8**···**9)** by NMR. The use of various substituted compounds **7a-7c** clarified the correlation between the Hammett substituent constant σY and Δ*G*, showing that the electron-donating substituent increases the interaction energy [16].

Yamada et al. [18] investigated the effect of substituents on the pyridinium ring by comparing the cation–π and π-π interaction energies with a molecular seesaw balance. This balance has advantages in the measurement of the interaction energies as follows: (1) There are only two distinct conformers, allowing the exclusion of undesirable conformational flexibility, and (2) as the two conformers show a high degree of symmetry, direct comparison of the conformers is possible without the need for any reference compounds. It was found that the presence of an electron-withdrawing substituent strengthens the interaction energy by reducing the electron density of the cation part, while an electron-donating substituent weakens the interaction. In the case of the Cl substituent, in particular, the cation–π interaction is enhanced by −3.09 kcal/mol compared to the π-π interaction (Table 2.3). The Δ*G* values significantly depend on solvent polarity. Solvents with large dielectric constants decrease the strength of the interaction (see Sect. 2.4.1 for a discussion of the solvent effect on interaction energy).

2.2.3 *Quaternary Ammonium Ions*

Ammonium, methylammonium, trimetylammonium, and tetramethyl ammonium are also known to form complexes with benzene, the interaction energies of which depend

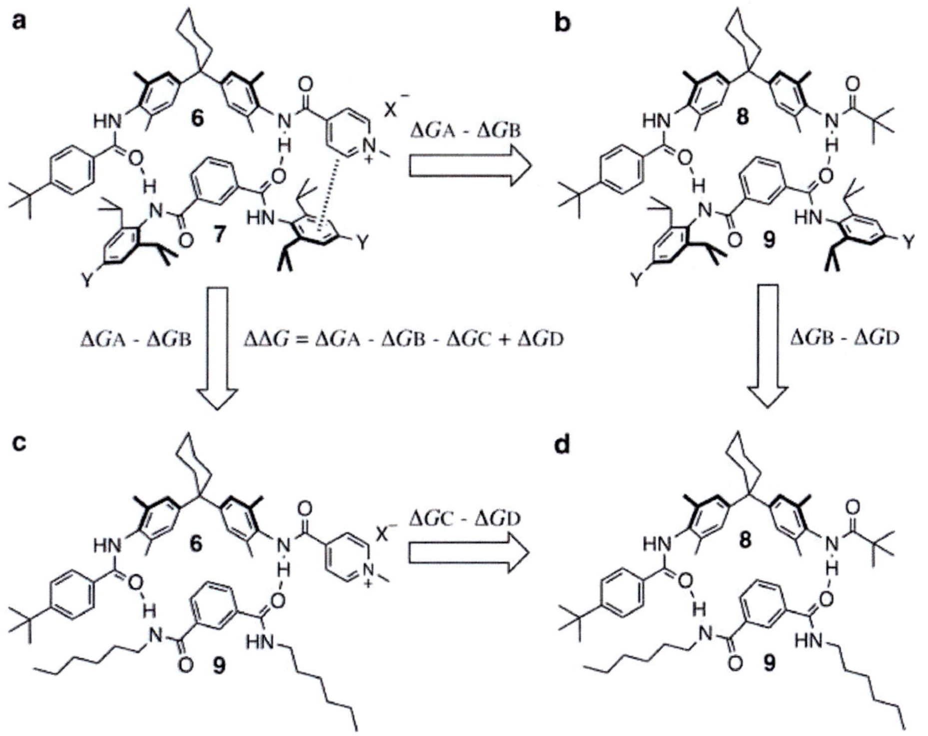

Fig. 2.8 Chemical double-mutant cycle for determining the magnitude of the cation–π interaction in complex **A**

Table 2.3 Evaluation of cation–π interactions using molecular seesaw balances

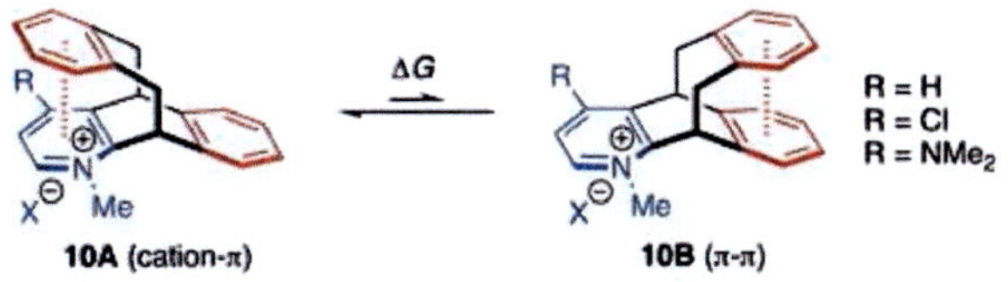

ΔG (kcal/mol)[a]				ε	1/ε
Solv	**10a** (R=H)	**10b** (R=Cl)	**10c**(R=NMe_2)		
$CDCl_3$	−1.47	−3.09	−1.36	4.8	0.208
CD_2Cl_2	−1.11	−1.80	−1.10	8.9	0.112
$(CD_3)_2CO$	−0.73	−0.86	−1.03	20.6	0.049
CD_3OD	−1.01	−1.3	−1.26	32.7	0.031
CD_3CN	−0.47	−0.63	−0.66	35.9	0.028
$(CD_3)_2SO$	−0.55	−0.62	−0.78	46.4	0.022

[a] Determined by ^{1}H NMR at 298 K

Table 2.4 Interaction energies for NR_4^+-π clusters[a]

Cluster	Energy	Cluster	Energy
NH_4^+/C_6H_6	−19.3	$CH_3NH_3^+/C_6H_6$	−18.8
NH_4^+/1,3,5-$C_6H_3(CH_3)_3$	−21.8	$CH_3NH_3^+/C_6H_{12}$	−11.6
NH_4^+/C_6H_5F	−14.4	$(CH_3)_3NH^+/C_6H_6$	−15.9
NH_4^+/C_2H_4	−10.0	$(CH_3)_4N^+/C_6H_6$	−9.4

[a] All energies are shown in kcal/mol

on their steric bulkiness. A smaller ammonium is effective for this interaction as shown in Table 2.4 [19]. In addition to the aromatic rings, alkene also serves as a π-component. While an electron-donating substituent on the benzene ring increases the interaction energy, an electron-withdrawing substituent decreases it.

While attraction between CH_4 and a π-plane is due to a CH-π interaction, the NH_4^+-π interaction is categorized as a cation–π interaction although their geometries are quite similar. Tsuzuki et al. [20] provided a detailed analysis of the interaction between ammonium and benzene using theoretical calculations. As shown in Table 2.5, the interaction energy is −19.3 kcal/mol, which is close to the value of the $K^+\cdots$ benzene interaction (Table 2.2). The contributions of electrostatic force and induction force are very large in this interaction. On the other hand, the interaction energy between ammonia and benzene is very small (−2.22 kcal/mol), with the major contribution being dispersion force. Based on these differences, the interaction between ammonium and benzene can be considered as a cation–π interaction rather than a NH-π interaction. The interaction between tetramethylammonium and benzene also has a large energy as in the case of ammonium. In this case also, the contributions of E_{es} and E_{ind} are large, which is consistent with the characteristic features of cation–π interactions. It is clear that the nature of the interaction is quite different from that of the CH-π interaction between methane and benzene, where the interaction energy is small (−1.45 kcal/mol) and the contribution of dispersion force is large.

In the formation of inclusion complexes of cyclophane **11** with **12** and **13**, **11** strongly encapsulates acetylcholine **12**, whereas little complexation is observed with neutral molecules **13** having a similar structure to **12** (Fig. 2.9) [21]. This strongly suggests that the positive charge of the ammonium plays a key role in the interaction.

Table 2.5 Interaction energies of benzene with cations and neutral molecules[a, b]

Complex	E_{total}	E_{es}	E_{ind}	E_{rep}	E_{corr}
C_6H_6/NH_3	−2.22	−1.01	–	1.14	−2.36
C_6H_6/CH_4	−1.45	0.25	–	1.10	−2.30
C_6H_6/NH_4^+	−19.30	−11.04	−13.58	9.58	−4.26
$C_6H_6/N(CH_3)_4^+$	−9.66	−5.27	−3.80	4.27	−4.85

[a] CCSD(T) level calculations, [b] In kcal/mol

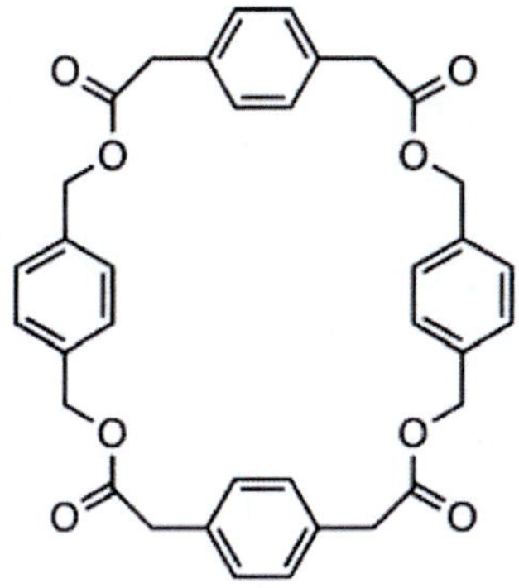

11

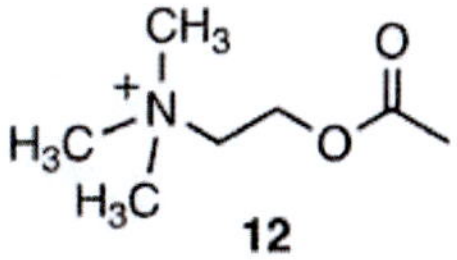

12

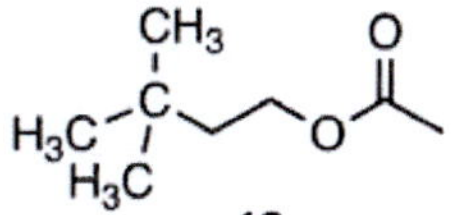

13

Fig. 2.9 A cyclophane host with cationic and neutral guest

Figure 2.10 shows examples of biological cation–π binding sites, the "aromatic box," of acetylcholine-binding protein and histone H3 protein [22]. Their aromatic residues form boxes, which recognize the quaternary ammonium of acetylcholine and the trimethyllysine side chain of histone protein, respectively, through cation–π interactions. This type of cation–π interaction is often observed in various biological systems as described in Chap. 3.

As quaternary ammonium ions have a three-dimensional structure, the distance between the central nitrogen atom and the aromatic ring is relatively distant compared to the planar pyridinium cation and the spherical alkali metal ion. However, the interaction energy is comparable to those ions. This can be explained by the fact that the positive charge is not localized on the N atom but distributed on the adjacent C atoms. As shown in Fig. 2.11, the calculations for ethyltrimethylammonium clarified

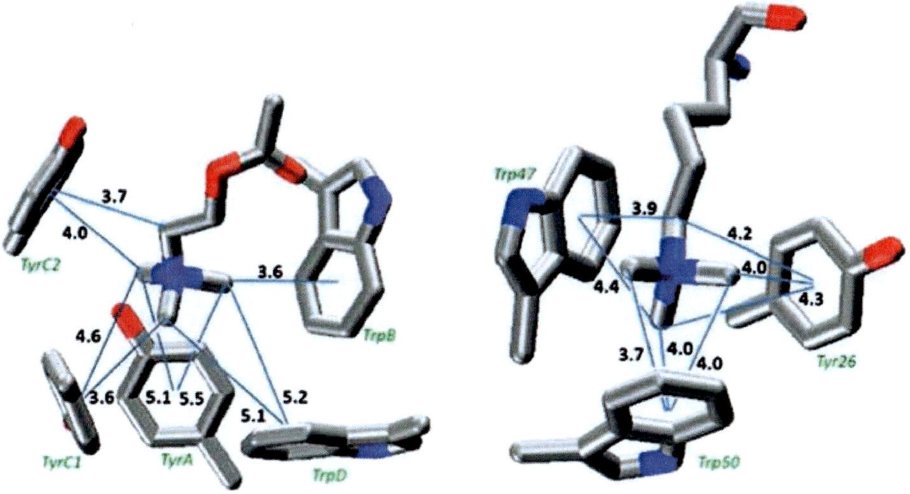

Fig. 2.10 "Aromatic boxes" of (left) acetylcholine-binding protein and (right) histone H3 protein. Reproduced from Ref. [22] with permission from the PCCP Owner Societies

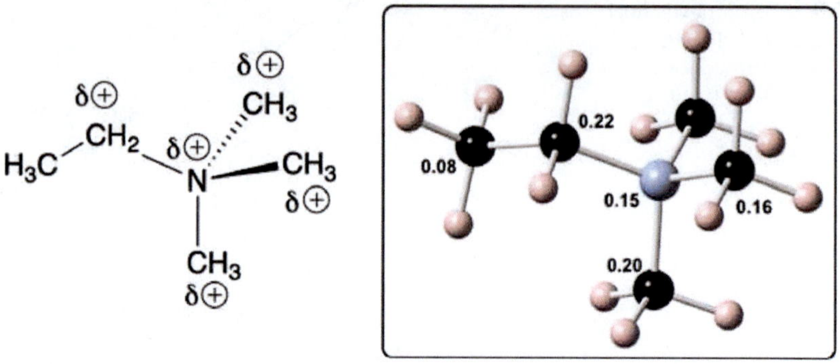

Fig. 2.11 Atomic charges obtained by ab initio calculations at the MP2/6-311G** level. Atomic charges with hydrogens were summed into C atoms. Reprinted with permission from [50]. Copyright 2018 American Chemical Society

that the methyl and ethyl groups are positively charged, which is the reason for the strong interaction between R_4N^+ and the π-system [23].

As will be discussed later (Chap. 3), tryptophan residues are known to play an important role in controlling the 3D structures of proteins through cation–π interactions. On the other hand, it is interesting to note that tryptophan itself takes on a unique conformation when it is protonated. The IR-IR-UV hole-burning spectroscopy experiments revealed the existence of two conformers **A** and **B** for protonated tryptophan (Fig. 2.12) [24]. The two conformers have closed conformations, with the ammonium moiety on the indole ring, and these are consistent with the two-lowest conformers obtained by ab initio calculations. In order to evaluate the binding energies between the ammonium and the indole moieties in these geometries, an analogous $NH_4^+\cdots$ indole complex as a model system was used for calculations. The interaction energy obtained at the RI-MP2/aug-cc-pVDZ level was 17.2 kcal/mol, in which the major sources of the attraction are the electrostatic and induction forces, suggesting contribution of a cation–π interaction.

2.2.4 *Guanidinium Ions*

Gallivan and Dougherty showed that there is a high probability of cation–π interactions between cationic and aromatic amino acid residues in a variety of proteins in the Protein Data Bank (PDB) [25]. A total of 593 proteins were examined, and, as shown in Table 2.6, there were a large number of pairs with stabilization energies greater than 2 kcal/mol. The cationic amino acid residues Arg and Lys interacted with the aromatic amino acid residues Phe, Tyr, and Trp at high probabilities. This indicates that the cation–π interactions of guanidinium and ammonium ions with benzene rings are commonly involved in the structure of proteins. As an example, a partial structure of the human growth hormone receptor is shown in Fig. 2.13 [26]. The cation-π interactions underlying the network structure of Trp(W)186 /Arg(R)211 /Phe(F)225

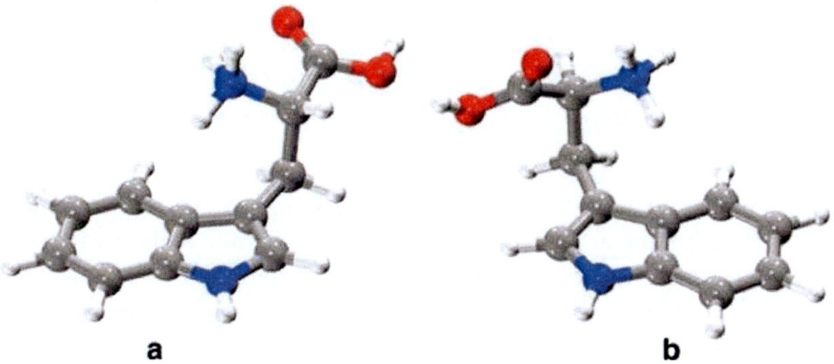

Fig. 2.12 Equilibrium structures of two lowest-energy L-TrpH^+ conformers, calculated at the CCSD/cc-pVDZ level of theory (silver = C, red = O, blue = N, white = H). Reprinted with permission from [24]. Copyright 2016 American Chemical Society

Table 2.6 Frequency of cation–π interactions within proteins

Amino acid pair	Cation–π	Percentage
Lys/Phe	285	14.5
Lys/Tyr	438	14.7
Lys/Trp	283	30.2
Arg/Phe	630	21.0
Arg/Tyr	749	20.6
Arg/Trp	609	40.4

/Arg(R)213 /Tyr(Y)222 residues were observed in the crystal structure, which are responsible for the formation of the receptor's 3D structure. This interacting system can be categorized as π-(cation–$\pi)_n$ system described in Sect. 2.5.2.2.

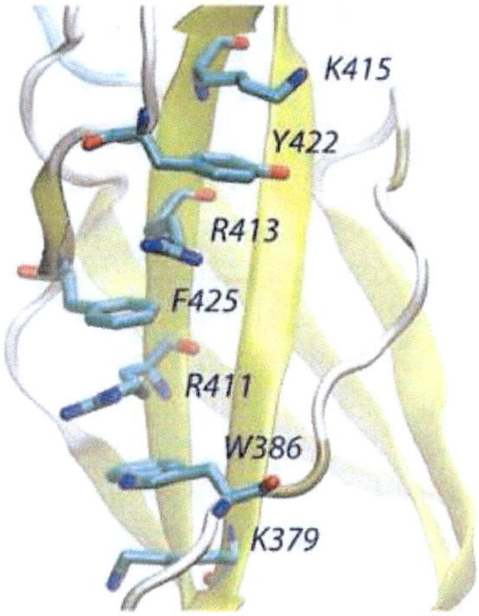

Fig. 2.13 Multiple cation–π-cation interactions identified in the human growth hormone receptor (PDB entry 1A22). Reproduced from Ref. [26] with permission from the PCCP Owner Societies

2.3 π-Systems

2.3.1 Aromatic Rings

The benzene ring is the most common π-face found in cation–π interactions. For example, the aromatic amino acid residues, phenylalanine and tyrosine, play a vital role in the formation of the three-dimensional structures of proteins. The size of the π-face, i.e., the degree of aromatic ring condensation, affects the interaction energy of the cation–π interaction [27] As the size of the π-plane increases, the interaction energy generally increases, as shown in Table 2.7. The interactions of Li^+ and Mg^{2+} with naphthalene result in an increase in the energies of −2.6 and −15.8 kcal/mol, respectively, compared to those with benzene. In the case of the anthracene$\cdots Li^+$ complex, the energies of the outer and inner rings (indicated in 1 and 2, respectively) are −40.2 kcal/mol and −39.6 kcal/mol, respectively, showing that the interaction energy of the outer ring is slightly larger than that of inner ring. Recently, the interaction with graphene, which possesses a very large π-plane, has attracted a lot of attention due to its various applications (see, Chap. 5).

As described above, the effect of aromatic substituents on the interaction energy has been experimentally clarified (Fig. 2.8). Table 2.8 shows the calculated total energies and electrostatic energies of arene$\cdots Na^+$ interactions. It is clear that the substitution of an electron-donating group increases the interaction energy, while an electron-withdrawing group decreases the interaction energy. A good correlation between E_{total} and E_{es} indicates that the aromatic substituents have a significant effect on E_{es} and that the E_{es} is a major contributor to the cation–π interactions [28].

2.3.2 Alkenes and Alkynes

Alkenes and alkynes also serve as π-components in cation–π interactions. The interaction energy between ethylene and ammonium has been experimentally determined to be about 10 kcal/mol [19]. On the other hand, the interaction energy between Li^+ and ethylene was estimated to be −19.7 kcal/mol, which is larger than that of ammonium. For acetylene, the stabilization energies with Li^+, Na^+, and K^+ have been estimated to be −20.20, −12.49, and −8.09 kcal/mol (NEO/MP2, BSSE corrected), respectively [29].

Interactions of cations with double bonds become stronger as the number of conjugations increases (Table 2.9). The interaction energy of Li^+ with butadiene is −26.4 kcal/mol, which is −6.7 kcal/mol larger than that with ethylene. As the conjugated system is further extended, it is further stabilized. A more pronounced effect was observed for the divalent ion Mg^{2+} [27]. For triene and tetraene, the stabilization effect differs between the outer and central bonds (indicated in 1 and 2, respectively). The complexes at the outer position of 18c(1) and 18d(2) are more

Table 2.7 MP2 interaction energies of the various aromatic-conjugated π-systems with Li^+ and Mg^{2+}

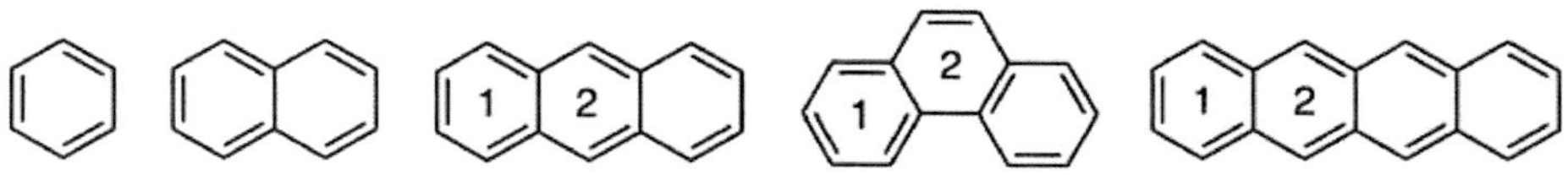

14a **14b** **14c** **14d** **14e**

π-system[a]	Interaction energy[b]	
	Li^+	Mg^{2+}
14a	−35.4	−108.7
14b	−38.0	−124.5
14c(1)	−40.2	−136.2
14c(2)	−39.6	−134.9
14d(1)	−41.2	−134.5
14d(2)	−39.3	−135.7
14e(1)	−41.8	−144.4
14e(2)	−41.4	−144.2

[a] The number in parentheses indicates the position where the cation is closest. [b] In kcal/mol

Table 2.8 Interaction energies for substituted benzene derivatives and Na^+

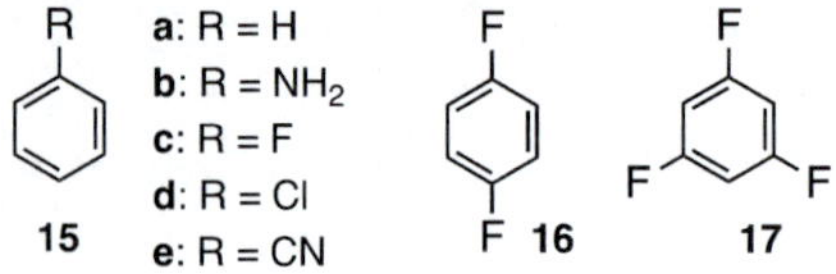

Complex	E_{total} [a]	E_{es} [a]
15a/Na^+	−27.1	−15.6
15b/Na^+	−31.8	−20.7
15c/Na^+	−22.0	−10.5
15d/Na^+	−21.5	−9.8
15e/Na^+	−15.7	−3.7
16/Na^+	−16.8	−5.2
17/Na^+	−12.4	−1.2

[a] In kcal/mol

stable than the complexes at the central position of 18c(2) and 18d(4). This trend is similar to the condensed aromatic systems described above (Table 2.7).

Table 2.9 MP2 interaction energies of the various acyclic-conjugated π-systems with Li^+ and Mg^{2+}

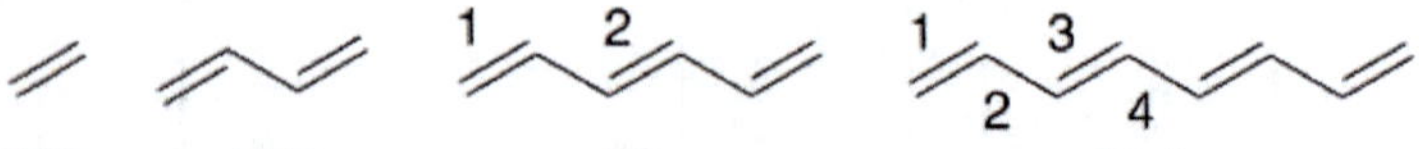

18a **18b** **18c** **18d**

π-System[a]	IE / kcal mol^{-1}	
	Li^+	Mg^{2+}
18a	−19.7	−68.8
18b	−26.4	−91.0
18c(1)	−29.2	−106.7
18c(2)	−26.6	−98.6
18d(2)	−31.5	−121.5
18d(4)	−31.1	−114.8

[a] The number in parentheses indicates the position where the cation is closest

2.3.3 *Carbonyl and Thiocarbonyl Groups*

The C=S and C=O groups also have π-surfaces that interact with cations in a manner similar to that of alkenes and alkynes. The strength of the interactions with cations is smaller than that observed for aromatic compounds. As these functional groups are widely found among organic compounds, the cation–π interactions that involve them are expected to be applied for various purposes as described in Chaps. 4 and 5.

Concrete examples for participation of these groups are shown in Fig. 2.14, which shows the X-ray crystal structures of nicotinic amide derivatives with thiazolidine-2-thione and oxazolidine-2-one groups [30]. A comparison between the nicotinic amide **19** and corresponding *N*-methyl salt **20** revealed the existence of C=S$\cdots$Py$^+$ interactions: the C=S$\cdots$Py$^+$ distance (3.53 Å) being shorter than the C=S$\cdots$Py$^+$ distance (3.88 Å). The geometrical difference between them is clearly shown in the superimposed structures. Similarly, an interaction between the C=O group and the pyridinium ring was observed in the nicotinic amide **22**. A comparison of **21** and **22** shows the C=O$\cdots$Py$^+$ distance (3.55 Å) to be a little shorter than that of C=O$\cdots$Py$^+$ (3.57 Å).

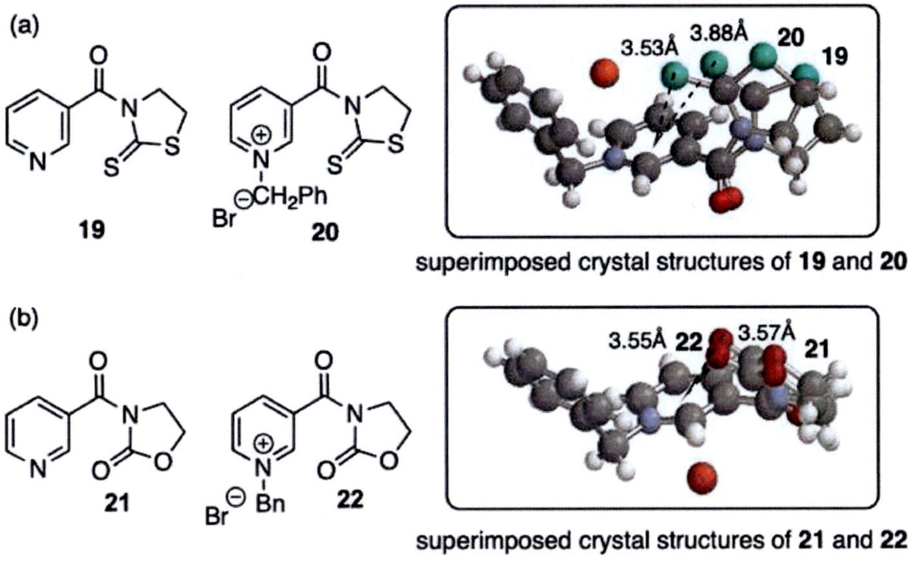

Fig. 2.14 Superimposed crystal structures for **a** **19** and **20** and **b** **21** and **22** with superimpositions. Reprinted with permission from [30]. Copyright 2004 American Chemical Society

2.4 Factors Affecting Interactions

2.4.1 *Solvent Effect*

The effect of a solvent on the interactions has been studied by theoretical calculations as well as experimental approaches. Table 2.10 shows the effects of the dielectric constant ε on the interaction of methylammonium with benzene and that of methylammonium with acetate ion for comparison. The interaction energy for the $MeNH_3^+ \cdots C_6H_6$ complex depends on the dielectric constant ε of the solvent, and the interaction energy decreases with an increase in the ε value. The interaction energies are −12.5 kcal/mol in the gas phase and −7.8 kcal/mol in carbon tetrachloride, while in water, it is slightly smaller at −5.5 kcal/mol [31].

On the other hand, in methylammonium acetate, the interaction energy between the cation and anion of the ion pair is greatly affected by the dielectric constant ε. In the liquid phase, the interaction energy is significantly lower than in the gas phase. In non-polar solvents, the ion-pair interaction is much stronger than the cation–π interaction. In contrast, in acetonitrile and water, the cation–π interaction energy is larger than that of the ion-pair interaction. This may result from the fact that a cation is still close to benzene ring and can overcome solvation by polar solvents. This shows that the cation–π interaction effectively works in a wide range of solvents and that cation–π interactions are not as sensitive to the solvent as are ion-pair interactions. The effect of the solvent on cation–π interactions has been investigated experimentally using a molecular balance. There is an almost linear relationship between the ΔG and $1/\varepsilon$ plots (see, Table 2.3).

Table 2.10 Interaction energies for N-methylammonium with benzene and acetate ions[a, b]

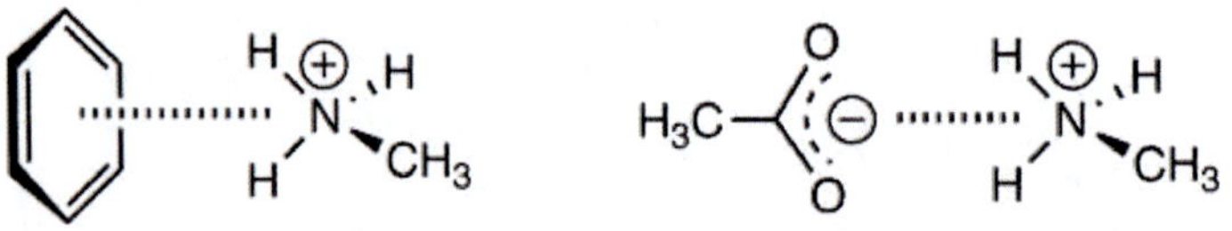

Solvent	Dielectric const ε	$C_6H_6/CH_3NH_3^+$	$CH_3CO_2^-/CH_3NH_3^+$
none	1.0	-12.5	−125.5
CCl_4	2.23	−7.8	−53.4
CH_3CO_2Et	5.99	−6.2	−19.7
CH_3CH_2OH	24.85	−5.6	−5.2
CH_3CN	37.5	−5.6	−3.8
H_2O	78.0	−5.5	−2.2

[a] SM5.42R/HF/6−31 + G*, [b] In kcal/mol

Table 2.11 ΔG values and ionic radii for **10**·MeX with a variety of anions in $CDCl_3$ and $(CD_3)_2CO$

Anion	r_d (pm)	ΔG (in $CDCl_3$)[a]	ΔG (in $(CD_3)_2CO$)[a]
Cl^-	181	−2.04	−0.76
Br^-	196	−1.78	−0.75
I^-	220	−1.47	−0.73
BF_4^-	228	−1.41	−0.75
$CH_3CO_2^-$	232	−1.19	−0.73
PF_6^-	245	−1.21	−0.75

[a] In kcal/mol

2.4.2 *Counter-Anion Effect*

In cation–π interactions, a cation always accompanies an anion. Due to the complex equilibrium of the π-component, cation, and anion in solution, it is difficult to elucidate the effect of the counter-anion on the interaction energy. Hunter et al. [17] solved this problem using the chemical double-mutant method with four complexes **A-D** consisting of four compounds **6–9** described in Sect. 2.2.1 (Fig. 2.8). By comparing the stabilization energies of the target complex **A**(6–7) with the other complexes **B**(8–7), **C**(6–9), and **D**(8–9), the contribution of the cation–π interaction alone in **A**(6–7) could be evaluated (Fig. 2.8). For the counter anions of BPh_4^-, PF_6^-, and I^-, the interaction energies were found to be within 2.5±0.4 kJ/mol, and there are no significant differences due to the paired anions. This indicates that although the counter anions affect the stability of the complex, they do not affect the magnitude of the interaction energy itself.

The effect of counter anions on intramolecular cation–π interactions was investigated using a molecular seesaw balance [18], and it was found that counter anions do affect the interactions (Table 2.11). It is clear that the effect of the anions on intramolecular cation–π interactions depends on the ionic radius of the anions in the non-polar solvent $CDCl_3$. In contrast, the size of anion does not affect the interaction energy in polar solvent (Fig. 2.15). This may be due to the formation of ion pairs with the cation in $CDCl_3$, which interferes with the intramolecular interactions, whereas the ion pairs would not be formed in $(CD_3)_2CO$. Although the reason for these results remains unclear, the steric nature of the counter anions as well as the H-bond of $PyMe^+$ with the anion may explain these results.

2.4.3 *Directional Dependence*

The cation–π interaction energy depends on the position of the cation relative to the π-surface. The interaction energies of benzene with ammonium and alkali metal cations have been calculated by Sherrill and coworkers [32] for various geometries, and the energies at $\theta = 0$, 30, 60, and 90° and $\phi = 0$ and 30° are given in Table 2.12.

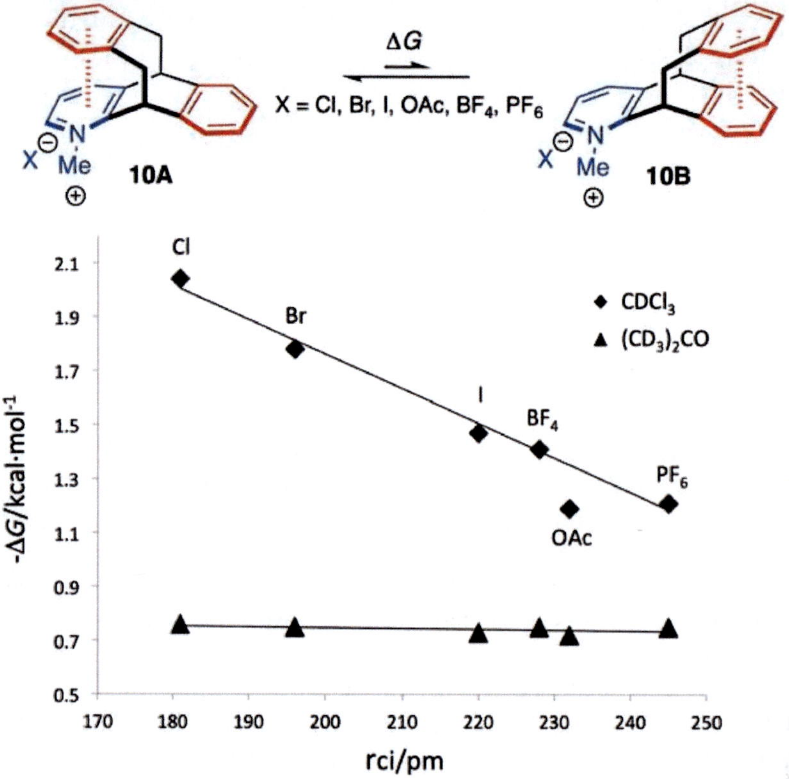

Fig. 2.15 Plots of ΔG vs r_{ci} for the molecular balance **10** in $CDCl_3$ ($r^2 = 0.991$ after exclusion of the ΔG value for an acetate ion) and in $(CD_3)_2CO$. Reprinted with permission from [18]. Copyright 2016 American Chemical Society

The most favorable geometry is when the cation is on the ring ($\theta = \phi = 0°$). The interaction energy was −35.8 kcal/mol for Li^+. The interaction energy decreased as θ increased from 0 to 30, 60, and 90°. These directional effects of cation–π interactions are attributable to the directionality of the electrostatic force arising from the charge distribution of a benzene ring as shown in Fig. 1.3.

Although the interaction energy decreases as the cation moves from the aromatic plane ($\theta = 0°$) to the side-on geometry ($\theta = 90°$), attraction is still observed in the side-on state (optimal R value is 1.33 Å for Li^+). This suggests that E_{es} would be positive in the side-on geometry due to repulsion between the cation and the hydrogen atom of the benzene ring, whereas E_{ind} is sufficiently negative to compensate for the positive E_{es} value. In the case of $\phi = 30°$, where the cation is located between two hydrogen atoms, the attraction with the benzene ring is larger than that in the case of $\phi = 0°$ as the repulsion between the cation and the hydrogen atoms of the benzene ring is reduced and R value is shortened, leading to enhancing E_{ind} value.

The interaction energy potentials for the five orientations of the benzene-pyridinium complexes (A-E) are shown in Fig. 2.16 [11]. The T-type complex C,

Table 2.12 Interaction energies for Li^+, Na^+, K^+, and NH_4^+ with benzene rings in various θ and ϕ

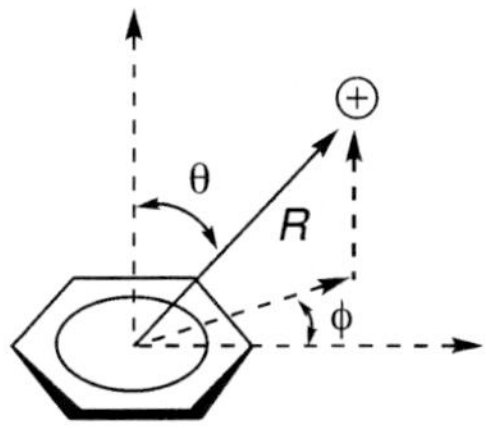

Cation	θ(°)	ϕ(°)	E (kcal/mol)	R(Å)
Li^+	0	0	−35.8	1.9
	30	0	−27.3	2.5
	60	0	−15.3	3.4
	90	0	−5.0	4.3
	30	30	−27.4	2.4
	60	30	−19.2	3.1
	90	30	−11.3	3.5
Na^+	0	0	−22.2	2.4
	90	0	−2.6	4.7
	90	30	−5.1	4.1
K^+	0	0	−16.5	2.9
	90	0	−1.5	5.2
	90	30	−3.0	4.5
NH_4^+	0	0	−16.4	3.1
	90	0	−1.7	5.3
	90	30	−3.4	4.6

in which the N−H of pyridinium is oriented toward benzene, has a much larger interaction energy than the other orientations due to the closer distances between the cation and the benzene ring. The T-type complex E, in which the C–H of benzene faces the center of the pyridinium, has a very small interaction energy compared to complex C. In the face-to-face complex A, the energy minimum is observed at the shortest distance compared to the other complexes.

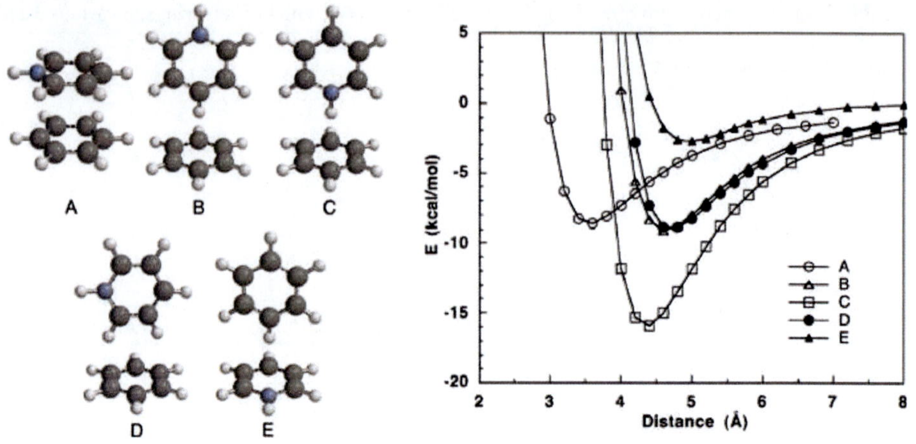

Fig. 2.16 MP2/cc-pVTZ level interaction energy potentials for benzene-pyridinium complexes. Intermolecular distance is the distance between the center of benzene and the midpoint of the nitrogen atom and 4-carbon atoms of the pyridinium. Reprinted with permission from [11]. Copyright 2007 American Chemical Society

2.5 Cooperation Among Non-covalent Interactions

2.5.1 Ternary Systems

In addition to cases in which one interaction works alone, there are cases where multiple interactions act cooperatively. In particular, two or more cooperating interactions are often observed in protein structures [33]. Figure 2.17 shows schematic illustrations of five types of interactions A–E, in which an additional cation, an anion, or a benzene ring is involved in the cation–π interaction system, respectively.

2.5.1.1 Cation–π–Cation (Type A) Systems

Cation–π–cation (Type A) systems are often found in protein structures. The analysis of proteins involving Tyr as an aromatic residue in the protein data set has clarified that interactions between Tyr and the ammonium group of Lys and the guanidinium moiety of Arg are common, with the Lys-Tyr-Arg motif detected in 26% of the X-Tyr-Y systems. In these systems, the same cationic amino acid residues (X-Tyr-X) were detected in 38%, with the most preferred being Lys-Tyr-Lys (19%) [26].

As an example of the three-component interaction system, the human growth hormone receptor (PDB entry 1A22) is shown in Fig. 2.13. Three successive cation–π-cation interactions, Lys-Trp-Arg, Arg-Phe-Arg, and Arg-Tyr-Lys, can be observed in the 3D structures. The interaction energies of these complexes are estimated to be −16.7, −1.0, and −7.8 kcal/mol [26], respectively, and support the idea that a variety of cation–π networks are utilized in the stabilization of three-dimensional

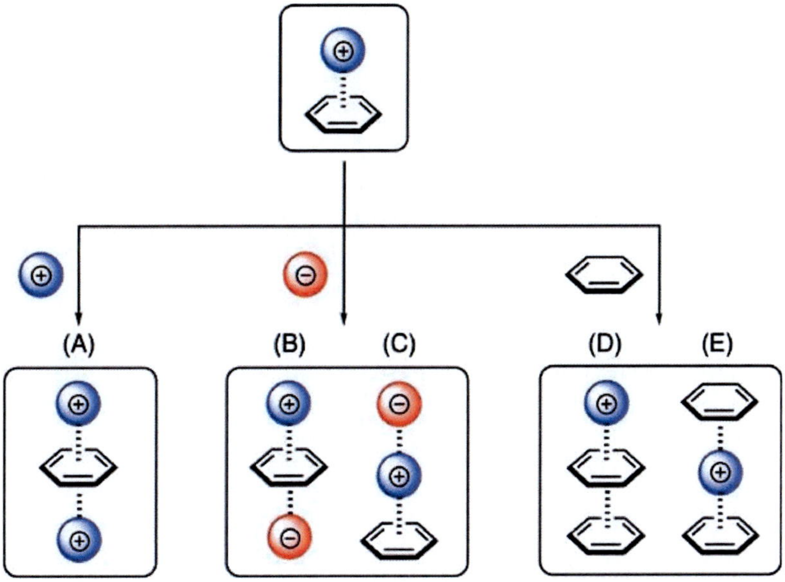

Fig. 2.17 Schematic representation of ternary complexes involving **a** Type A (cation–π–cation), **b** Type B (cation–π–anion), **c** Type C (anion–cation–π), **d** Type D (cation–π–π), and **e** Type E (π–cation–π) motifs

structures. This type of system involves not only cation–π–cation (Type A), but also π–cation–π (Type E) systems as discussed later (see, Sect. 2.5.1.3). Both contribute to the stabilization of the 3D structure of the protein. Therefore, it is more appropriate to classify this example as a π-(cation–π)$_n$ system (see, Sect. 2.5.2.2).

Cation–π interactions are generally accompanied by anions due to the charge balance. When the aromatic rings are electron deficient, anions are known to interact with them instead of cations, which are named as "anion-π" interactions [34, 35].

2.5.1.2 Cation–π–Anion (Type B) and Anion–Cation–π (Type C) Systems

Table 2.13 shows the calculated energies and equilibrium distances for cation–π–anion (X^+-π-Y^-) (Type B) ternary systems at the MP2 level. This also lists the interaction energy $E_{(Na^+-X^-)}$ for the ion pair Na^+-X^-, calculated on the distances of both ions obtained for each ternary complex. It can be seen that the stabilization energy of the ternary complex with benzene is greater than that in the complex with hexafluorobenzene. The interaction energies of the ternary complexes are $-12 \sim -20$ kcal/mol stronger than those of the ion-pair complexes, indicating that the ion-pair complexes are stabilized by forming cation–π-anion complexes with the benzene and hexafluorobenzene rings [36].

Calculations for the anion-π-Py^+ (Type B) and anion-Py^+-π (Type C) ternary systems have also been performed [37]. A comparison of the two systems shows

Table 2.13 Interaction energies for ternary complexes(X^--π-Na^+) and for the Na^+-X^- ion pair at a fixed distance

Complex(X^--π-Na^+)	E_{BSSE}[a,b]	$E_{Na\text{-}X(BSSE)}$[a,b]	$Re_{(X^- - \pi)}$[c]	$Re_{(Na^+ - \pi)}$[c]
$F^-\cdots C_6H_6\cdots Na^+$	−92.42	−71.71	2.482	2.280
$Cl^-\cdots C_6H_6\cdots Na^+$	−84.48	−63.18	3.049	2.304
$Br^-\cdots C_6H_6\cdots Na^+$	−83.61	−62.04	3.157	2.313
$F^-\cdots C_6F_6\cdots Na^+$	−88.42	−71.33	2.286	2.437
$Cl^-\cdots C_6F_6\cdots Na^+$	−75.63	−63.58	2.835	2.488
$Br^-\cdots C_6F_6\cdots Na^+$	−74.04	−62.81	2.913	2.495

[a] In kcal/mol. [b] The binding energies (*E*) were calculated with the basis set superposition error (BSSE) by using the Boys–Bernardi counterpoise technique. [c] In Å

that when the Py^+ and anion are on the same side of the π-face (Type C), they are anticooperative. On the other hand, when the anion and the Py^+ are on the opposite sides of the π-face (Type B), the system is cooperative, in a similar manner to the case of Na^+. It has also been shown that the presence of an anion near the π-system causes a large redistribution of the charge density of the π-electrons, which enhances the induction force.

2.5.1.3 Cation–π–π (Type D) and π–Cation–π (Type E) Systems

Table 2.14 shows examples of Types D and E systems [38]. When an additional benzene interacts with a complex of a metal ion and benzene (BM), two types of complexes, D and E, are formed; the metal ion of D is on the benzene dimer and that of E is sandwiched by the two benzene molecules.

In both complexes, the interaction energies for all monovalent ions are larger than those for BM complexes, and a significant stabilization can be observed, particularly when the metal ion is divalent. This suggests that the interaction with benzene strengthens the cation–π interaction and, at the same time, the metal ion enhances the π-π interaction between the benzenes. This is evident from the changes in the distances R_1 and R_2. For example, in the case of complexes with Li^+, R_1 (1.932 Å) in BM shrinks to 1.913 Å in complex D, and R_2 is shortened from 3.768 Å, which is the general distance of the benzene dimer, to 3.548 Å.

As the complex E contains two cation–π interactions, the interaction energy is much larger than that for complex BM. The R_1 of complex E (M=Li) is 1.977 Å, which is longer than those of BM (1.932 Å) and complex D (1.913 Å). The longer R_1 of complex E in comparison to that of complex D may be due to repulsion of the dipole moments between the benzene rings. In aqueous solution, it has been reported that the K^+ ion prefers to form a 2:1 sandwich-type complex rather than a 1:1 cation–π complex [39]. Alkali metal ions are strongly bound to benzene rings in the gas phase, and their affinity for benzene follows the expected electrostatic trend ($Li^+ > Na^+ > K^+ > Rb^+$) as shown in Sect. 2.1. Dougherty et al., through simulations in an aqueous

Table 2.14 Interaction energies of a metal ion with benzene monomers and dimers. Reprinted with permission from [38]. Copyright 2006 American Chemical Society

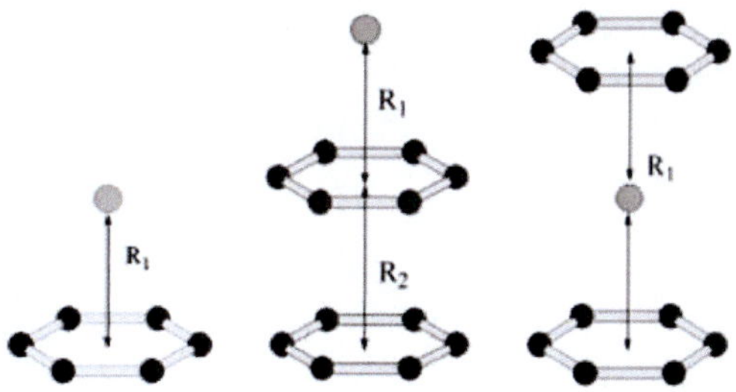

Cation	Complex BM[a]	Complex D[a]	Complex E[a]
Li^+	−34.13	−38.76	−57.88
Na^+	−20.26	−23.83	−35.57
K^+	−16.64	−19.81	−28.78
Mg^{2+}	−105.78	−120.05	−179.08
Ca^{2+}	−71.57	−82.66	−125.49

[a] In kcal/mol

environment, found that K^+ forms complexes more readily than does Na^+. More surprisingly, in the benzene···M^+···benzene complex, this order of preference is reversed, with K^+ taking precedence over all other ions (Fig. 2.18). This may be due to differences in solvation energies among metal ions; i.e., as the solvation energy of K^+ with water is weaker than that of Li^+ due to larger ion radius, desolvation loss of K^+ in the complex formation is less than in the case of Li^+ and, therefore, K^+ complex formation is preferred.

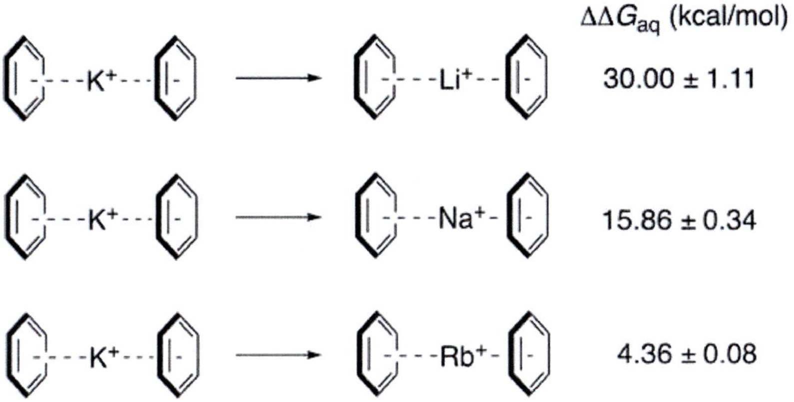

Fig. 2.18 $\Delta\Delta G_{aq}$ values for replacing K^+ with Li^+, Na^+, and Rb^+ in the benzene-K^+-benzene complex

The phenomenon underlying the preferential selection of potassium ions over all other ions in 2:1 benzene:ion complexes in an aqueous environment is analogous to voltage-gated potassium channels. As there are several conserved aromatic residues in the pore region of such potassium channels (see, Sect. 3.5), these results suggest that cation–π interactions may contribute to the ion selectivity of the potassium ion channels. It has also been reported that π–cation–π complexes are preferred in interactions with ammonium ions, in which the NH_4^+–π and π–π interactions act synergistically [40].

2.5.2 *Multiple Systems*

2.5.2.1 Cation-$(\pi)_n$-π Systems

It should be noted that Type D system (cation–π–π) can be expanded to multicomponent systems. Frontera and coworkers [41] reported synergistic effects between cation–π and π–π interactions in cation-$(\pi)_n$-π systems. The presence of a cation interacting with the π-cloud of an aromatic ring promotes stacking interactions with additional aromatic rings. This effect is extended with up to five aromatic ring stacking as shown in Fig. 2.19. The equilibrium distance of the lithium cation to the ring centroid (R_1) progressively shortens as the number of benzene rings increases (from 1.914 (n = 0) to 1.875 Å (n = 4)). The same effect is observed for the equilibrium distance of the π-π interaction (R_2) (from 3.591 (n = 1) to 3.545 Å (n = 4)), indicating that the subsequent addition of benzene rings also strengthens the π–π interaction. The synergistic energies increased from −4.1 to −6.0 kcal/mol for n = 1 to 4.

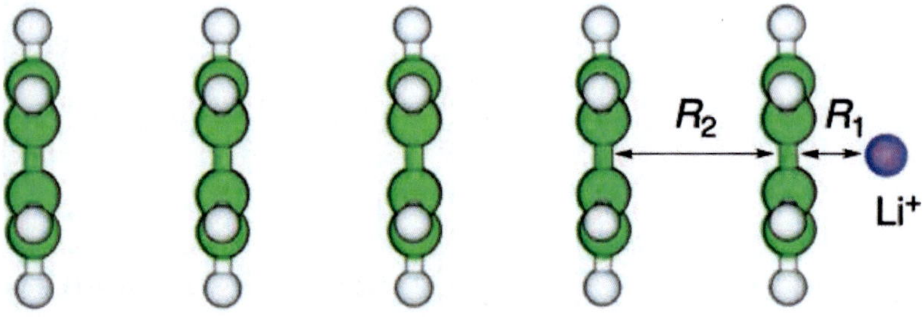

Fig. 2.19 Optimized structure of the Li^+-$(\pi)_4$-π complex calculated at the RI-MP2/6−31++G** level. Reprinted with permission from [41]. Copyright 2006 American Chemical Society

2.5.2.2 π–(Cation–π)$_n$ Systems

A combination of Type A and Type E can form π–(cation–π)$_n$ systems. Examples of this are observed in the crystal structures of proteins as well as small molecules. As mentioned in Sect. 2.5.1.1, the crystal structure of human growth hormone receptor shown in Fig. 2.13 contains both cation–π–cation and π–cation–π systems. Therefore, it is more appropriate to classify this example as a π–(cation–π)$_n$ system.

Figure 2.20a shows a crystal packing diagram of compound **5**, indicating not only an intramolecular cation–π interaction as shown in Fig. 2.7, but also an intermolecular cation–π interaction forming a columnar motif [15]. The separation between the pyridinium and benzene rings is 3.780 Å and 3.779 Å, respectively, indicating the existence of a cation–π interaction network.

A similar columnar motif was observed in compound **20** as described in Fig. 2.20b. In addition to the intramolecular interaction, intermolecular interactions between the thiocarbonyl and the pyridinium ring play a key role in the construction of the columnar motif [30]. The separation between the pyridinium ring and the S atom of the thiocarbonyl group is 3.53 Å and 3.90 Å, respectively.

Figure 2.21a shows the crystal structure of compound **23**·2H_2O [43]. The molecules of **23** are arranged in a head-to-tail and face-to-face manner to form a column structure through π-(cation–π)$_n$ interactions between the pyridinium and benzene rings. The four columns form a channel, which the chloride ions and water molecules occupy. The distances between the double bonds are 3.797 Å and 4.070 Å. These features are responsible for the high photo-reactivity as well as the synHT

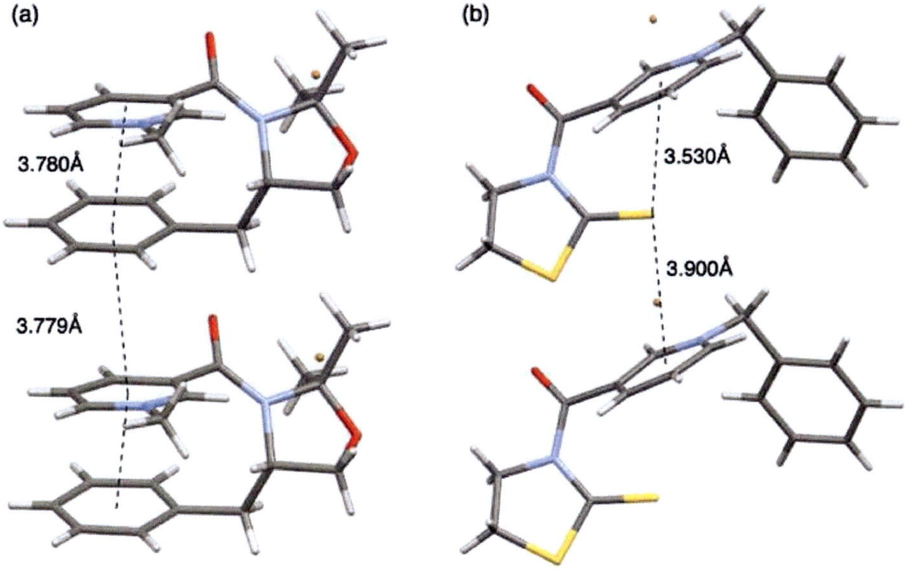

Fig. 2.20 Packing diagrams of **a** compound **5** and **b** compound **20**. Reprinted from [42], Copyright 2020, with permission from Elsevier

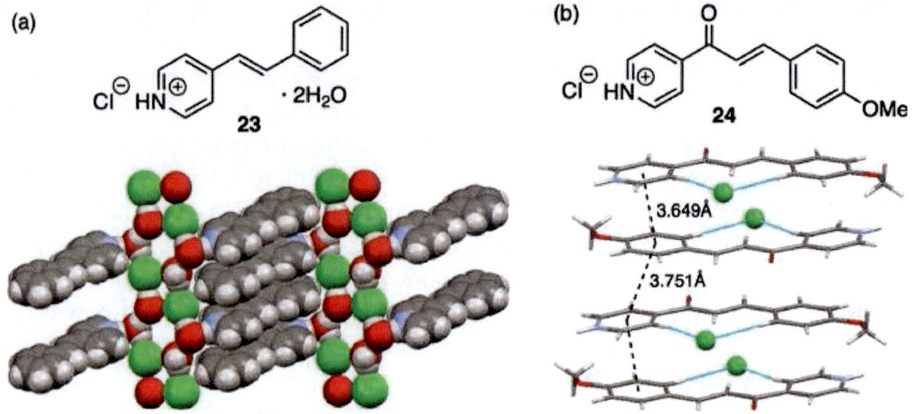

Fig. 2.21 Packing diagrams of **a** compound **23**·$2H_2O$ and **b** compound **24**. Reprinted from [42], Copyright 2020, with permission from Elsevier

dimer formation (see, Sect. 4.4). It should be noted that the corresponding neutral molecules in the crystal are oriented in a head-to-head manner, with the distance between the double bonds being 7.14 Å, leading to their photostability.

Figure 2.21b shows the packing diagram of **24** [44]. The molecules are also oriented in a head-to-tail manner, forming an infinite column structure through cation–π interactions. The distances between the centroids of the pyridinium and benzene rings are 3.649 and 3.751 Å. These structural features are thought to be due to the stabilization of the crystal packing structure through the cooperative effects of cation–π interactions.(Fig. 2.21).

2.5.3 Cooperation with Hydrogen Bonds

Pyrrole and related compounds have a N–H moiety in their ring systems, which can form a hydrogen bond with a proton acceptor. Calculations (MP2/6–31++G**) for complexes of pyrrole (**25**) involving both a cation–π interaction and a hydrogen bond indicate the presence of synergistic effects.

The geometries and energies for the complexes A-C of pyrrole (**25**) with H_2O, Na^+, and K^+ were calculated and are summarized in Table 2.15 [45]. The equilibria for the hydrogen bond distance (R_{HB}) and cation–π distance (R_e) for these complexes are also shown in the Table. The distance R_e for the cation–π-σ complexes **C** is shorter than that for the binary complexes **B**, indicating that the presence of the hydrogen bond enhances the cation–π interaction. On the other hand, the equilibrium distance R_{HB} of the σ interaction in complexes **C** is also shorter than that for the H-bond complex **A**, indicating that the presence of the cation–π interaction enhances the H-bond interaction.

Table 2.15 Binding and synergistic energies and equilibrium distances for complexes A–C[a]

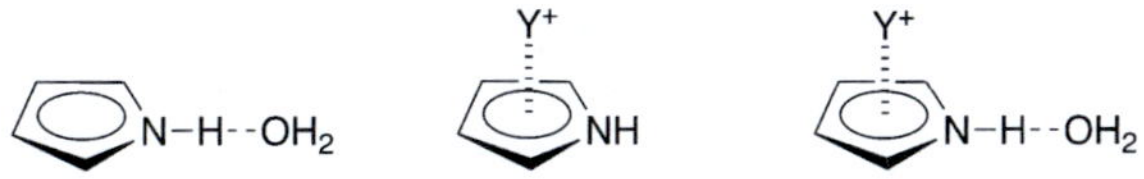

A ($\mathbf{25}\cdots H_2O$) B ($\mathbf{25}\cdots Y^+$) C ($Y^+\cdots\mathbf{25}\cdots H_2O$)

Complex	E_{BSSE} [b]	E_{syn} [b]	R_{HB} [b]	R_e [b]
A(**25**—H_2O)	−5.07	–	1.962	–
B(Na^+—**25**)	−24.26	–	–	2.489
B(K^+—**25**)	−17.91	–	–	2.880
C(Na^+—**25**—H_2O)	−35.97	−6.64	1.832	2.446
C(K^+—**25**—H_2O)	−28.75	−5.77	1.848	2.832

[a] Calculated at the MP2/6−31++G** level of theory. [b] In kcal/mol

Table 2.15 shows the calculated values of the synergistic energies (E_{syn}), which are the differences between the binding energies for the ternary complexes **C** and those for the associated H-bond and cation–π complexes **A** and **B**, respectively. The E_{syn} values for the complexes **C** with Na^+ and K^+ are −6.64 and −5.77 kcal/mol, respectively, strongly suggesting the contribution of a synergistic effect between the two interactions. Not only does E_{syn} include a contribution from the Coulomb interaction between the cation and water, but the influence of dielectric polarization due to the electric field between the cation and water (multibody effect) is also thought to contribute.

It is known that both H-bond and cation–π interactions are found between nucleobases and amino acid side chains. As shown in Fig. 2.22, a positively or partially positively charged amino acid side chain forms a hydrogen bond with a nucleobase and interacts with another nucleobase through a cation–π interaction. The repetition of these units forms a staircase motif.

A survey of 52 high-resolution structures of protein/DNA complexes revealed the presence of such motifs in a majority of complexes [46]. Figure 2.22 shows the longest quadruplex staircase observed in zinc finger 1MEY. It consists of five consecutive nucleobases, C-A-G-A-A, and four positively or partially positively charged amino acid residues, Gln-Lys-Asn-Gln, forming an H-bond and cation–π interaction network.

The interaction energies of NH_4^+ with neutral amino acids were investigated by Wu and McMahon [47]. A plot of the binding energies of five aliphatic amino acids, Gly, Ala, Val, Leu, and Ile, shows that there is a very good linear relationship between the binding energies and proton affinities (PAs), indicating the existence of a single interaction between the cation and the amino acid (Fig. 2.23a). However, three aromatic amino acids are exceptions to this trend. The optimized structure of protonated Tyr and ammonia suggested the a cation–π interaction between the

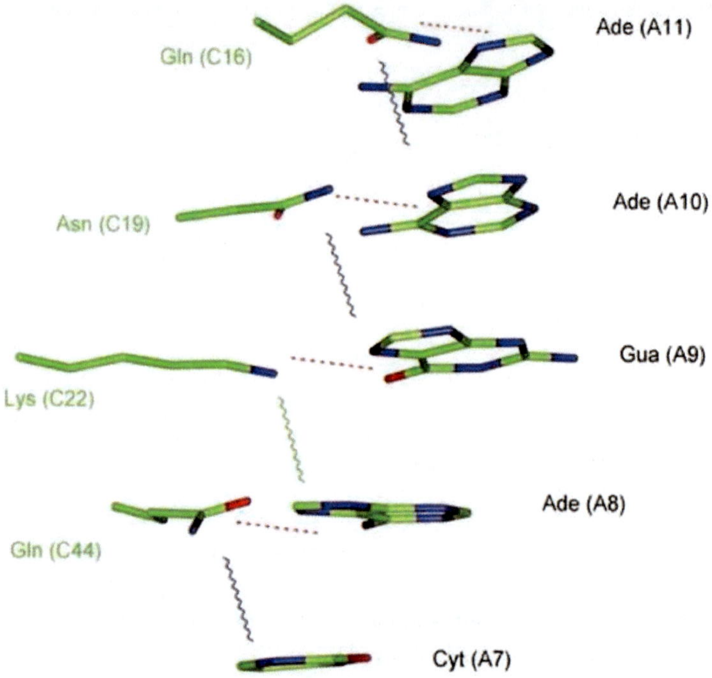

Fig. 2.22 A multiple stair motif observed in the Protein–DNA complex of the zinc finger 1MEY. Reprinted from [46], Copyright 2002, with permission from Elsevier

ammonium ion and the aromatic ring of the amino acid, as shown in the Fig. 2.23b. The deviation from the linear trend allows the degree of enhancement resulting from the cation–π interaction for the NH_4^+ clusters of Phe, Tyr, or Trp to be concluded to be −3.3, −3.7, or −3.6 kcal/mol, respectively.

This type of interaction can be categorized as a chelate effect. Ion-pair formation brings the ammonium ion close to the benzene ring of the amino acid, assisting the cation–π interaction without any loss of entropy [32].

2.5.4 Cooperation with Halogen Bonds

Cooperation effects between cation–π interactions and halogen bonds have also been reported. The effect of halogen bonding on cation–π interactions has been investigated by the MP2 calculations of $M^+\cdots ArX$ (A) and $ArX\cdots B$ (B) dyads and $M\cdots ArX\cdots B$ ternary complexes (C). Their geometric and energy data are given in Table 2.16 [48]. A comparison of the distances R_{cp} between the cation and the benzene ring shows that the distances for the ternary complexes (C) are shorter than those for the corresponding binary complexes (A). These findings clearly show that the presence of halogen bonds enhances the cation–π interactions.

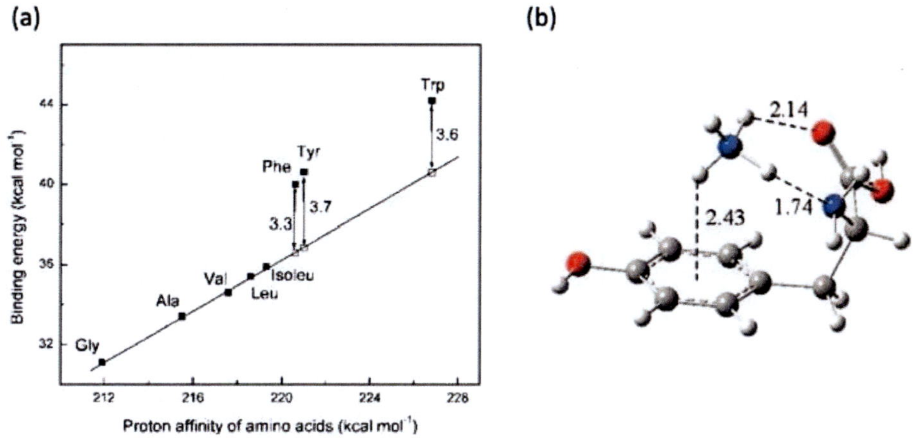

Fig. 2.23 **a** Binding energies with NH_4^+ vs. proton affinity of amino acids and **b** the optimized structure of protonated Tyr and ammonia obtained by B3LYP/6–31+G(d) Reprinted with permission from [47]. Copyright 2008 American Chemical Society

Table 2.16 Interaction energies, synergistic energies, and intermolecular distances for complexes A–C

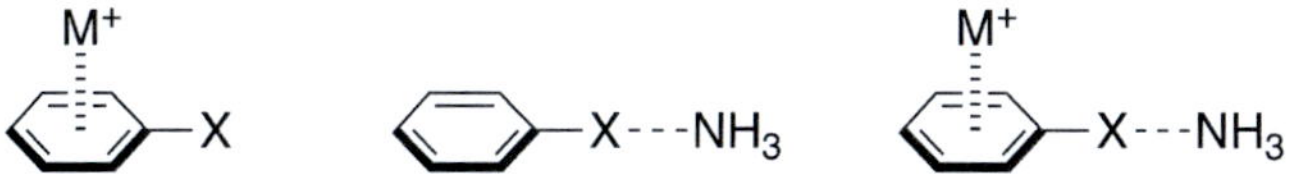

A ($M^+\cdots ArX$) B ($ArX\cdots NH_3$) C ($M^+\cdots ArX\cdots NH_3$)

Complex	ΔE (kcal/mol)	E_{syn} (kcal/mol)	R_{MAr}(Å)	R_{XB} (Å)
A($Na^+\cdots C_6H_5I$)	−19.61	–	2.507	–
A($Mg^{2+}\cdots C_6H_5I$)	−112.64	–	1.981	–
A($Na^+\cdots C_6H_5Br$)	−18.80	–	2.514	–
B($C_6H_5I\cdots NH_3$)	−3.03	–	–	3.106
B($C_6H_5Br\cdots NH_3$)	−1.57	–	–	3.102
C($Na^+\cdots C_6H_5I\cdots NH_3$)	−28.84	−6.20	2.470	2.877
C($Mg^{2+}\cdots C_6H_5I\cdots NH_3$)	−135.41	−19.74	1.937	2.574
C($Na^+\cdots C_6H_5Br\cdots NH_3$)	−25.36	−4.99	2.483	2.906

On the other hand, the halogen bond distances R_{XB} of the M···ArX···B (C) ternary complexes are shorter than those of the binary complexes (B). In $M^+\cdots C_6H_5I\cdots NH_3$ complexes, when Na^+ is replaced by Mg^{2+}, the halogen bond (ArX··· B) distance is significantly reduced from 3.106 Å to 2.574 Å. These results prove that the cation–π interaction strengthens the halogen bond. The fact that the calculated synergy energies in complex C lie between −4.99 and −19.74 kcal/mol clearly shows the

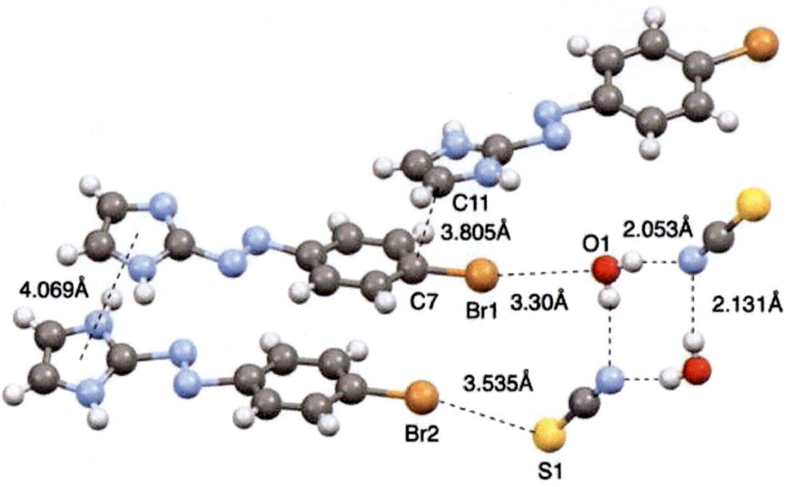

Fig. 2.24 Crystal structure of azoimidazolium thiocyanate

cooperation of the two interactions in strengthening each other. A comparison of the effects of Na^+ and Mg^{2+} on E_{syn} indicates effective stabilization by Mg^{2+} than Na^+.

This energy stabilization effect in triads can be explained by the charge transfer from the nitrogen atom to the benzene ring through the halogen bond. As the halogen bond is strengthened, the electron density of the aromatic ring increases, leading to strengthening of the cation–π interaction.

As an example of the synergistic effect of a cation–π interaction with a halogen bond, the crystal structure of the aryl azoimidazolium thiocyanate $[C_9H_7N_4Br]\cdot[C_9H_8N_4Br]^+\cdot[SCN]^-\cdot[H_2O]$ is shown in Fig. 2.24 [49]. This crystal is a hydrate salt formed from two aryl azoimidazoles and thiocyanate, with one of the imidazole rings protonated.

The bromine atom of the bromobenzene moiety forms a halogen bond with a water molecule, and the benzene ring approaches the protonated imidazolium ring of the other molecule through a cation–π interaction. The interatomic distance between the C11 of the imidazolium ring and the C7 of the bromobenzene is 3.805 Å, suggesting the existence of a cation–π interaction. The halogen bond Br1···O1 has a distance of 3.30 Å, which is shorter than the sum of the van der Waals radii (Br: 1.85 A; O: 1.52 A), and the C7-Br1···O1 angle is 167.95 Å, suggesting a moderately strong halogen interaction. These cation–π interactions and the halogen bond are thought to synergistically cooperate as described above.

Other interactions are also involved in the stabilization of the crystal structure, such as a cation–π interaction between the imidazolium and imidazole rings (4.069 Å for centroids), a halogen bond between the bromine atom Br2 and the sulfur atom of the thiocyanate (SCN) anion (3.535 Å for Br2···S1), and hydrogen bonds between the nitrogen atom of the thiocyanate ion and the two water molecules (2.053 and 2.131 Å).

These effects are very important in biology as there are a variety of heteroaromatic moieties in biological systems. Therefore, understanding the synergistic effects of the two interactions can help our understanding of the biological processes as well as of the structures of supramolecules and organic crystals. Furthermore, it is expected to play an important role in the fields of synthetic organic chemistry and materials chemistry in terms of molecular design and elucidation of reaction mechanisms [50].

References

1. Ma JC, Dougherty DA (1997) The cation–π Interaction. Chem Rev 97:1303–1324
2. Sunner J, Nishizawa K, Kebarle K (1981) Ion-solvent molecule interactions in the gas phase. The potassium ion and benzene. J Phys Chem 85:1814–1820
3. Tsuzuki S, Yoshida M, Uchimaru T, Mikami M (2001) The origin of the cation/π interaction: the significant importance of the induction in Li^+ and Na^+ complexes. J Phys Chem A 105:769–773
4. Tsuzuki S, Uchimaru T, Mikami M (2003) Is the cation/π interaction in alkaline-earth-metal dication/benzene complexes a covalent interaction? J Phys Chem A 107:10414–10418
5. Reddy AS, Zipse H, Sastry GN (2007) Cation–π interactions of bare and coordinatively saturated metal ions: contrasting structural and energetic characteristics. J Phys Chem B 111:11546–11553
6. Meadows ES, De Wall SL, Barbour LJ, Gokel GW (2001) Alkali metal cation−π interactions observed by using a lariat ether model system. J Am Chem Soc 123:3092–3107
7. Hu J, Barbour LJ, Gokel GW (2001) Solid state evidence for π-complexation of sodium cation by carbon-carbon double bonds. Chem Comm:1858–1859.
8. Hu J, Barbour LJ, Gokel GW (2001) Solid-state evidence for π-complexation of sodium and potassium cations by carbon−carbon triple bonds. J Am Chem Soc 123:9486–9487
9. Zhang S-L, Liu L, Fu Y, Guo Q-X (2005) Cation–π interactions of Cu^+. J. Mol. Stru: THEOCHEM 757:37–46
10. Stauffer DA, Barrans RE Jr, Dougherty DA (1990) Biomimetic catalysis of an S_N2 reaction resulting from a novel form of transition-state stabilization. Angew Chem Int Ed Engl 29:915–918
11. Tsuzuki S, Mikami M, Yamada S (2007) Origin of attraction, magnitude, and directionality of interactions in benzene complexes with pyridinium cations. J Am Chem Soc 129:8656–8662
12. Dewar MJS, Thompson CC Jr (1966) π-Molecular complexes–III. Tetrahedron 22:97–114
13. Bentley MD, Dewar MJS (1967) Some π-molecular complexes showing no charge transfer spectra formed by p-toluene slufonic esters. Tetrahedron Lett 8:5043–5047
14. Hunter CA (2004) Quantifying intermolecular interactions: guidelines for the molecular recognition toolbox. Angew Chem Int Ed 43:5310–5324
15. Yamada S, Morita C (2002) Face-selective addition to a cation−π complex of a pyridinium salt: synthesis of chiral 1,4-dihydropyridines. J Am Chem Soc 124:8184–8185
16. Hunter CA, Low CM, Rotger C, Vinter JG, Zonta C (2002) Substituent effects on cation-π interactions: a quantitative study. Proc Natl Acad Sci USA 99(8):4873–4876
17. Hunter CA, Low CMR, Rotger C, Vinter JG, Zonta C (2003) The role of the counteranion in the cation–π interaction. Chem Comm:834−835.
18. Yamada S, Yamamoto N, Takamori E (2016) Synthesis of molecular seesaw balances and the evaluation of pyridinium−π interactions. J Org Chem 81:11819–11830
19. Deakyne CA, Meot-Ner (Mautner) M (1985) Unconventional ionic hydrogen bonds. 2. $NH^+\cdots\pi$ complexes of onium ions with olefins and benzene derivatives. J Am Chem Soc 107:474−479.
20. Tsuzuki S, Fujii A (2008) Nature and physical origin of CH/π interaction: significant difference from conventional hydrogen bonds. Phys Chem Chem Phys 10:2584–2594

21. Roelens S, Torriti R (1998) Binding of acetylcholine and quaternary ammonium cations to macrocyclic and acyclic "phane" esters. Evaluation of the cation−π primary interaction through adaptive aromatic hosts. J Am Chem Soc 120:12443–12452
22. Davis MR, Dougherty DA (2015) Cation–π interactions: computational analyses of the aromatic box motif and the fluorination strategy for experimental evaluation. Phys Chem Chem Phys 17:29262–29270
23. Tsuzuki S, Tokuda H, Hayamizu K, Watanabe M (2005) Magnitude and directionality of interaction in ion pairs of ionic liquids: relationship with ionic conductivity. J Phys Chem B 109:16474–16481
24. Pereverzev AY, Cheng X, Nagornova NS, Reese DL, Steele RP, Boyarkin OV (2016) Vibrational signatures of conformer-specific intramolecular interactions in protonated tryptophan. J Phys Chem A 120(28):5598–5608
25. Gallivan JP, Dougherty DA (1999) Cation–π interactions in structural biology. Proc Natl Acad Sci USA 96:9459–9464
26. Pinheiro S, Soteras I, Gelpi JL, Dehez F, Chipot C, Luque FJ et al (2017) Structural and energetic study of cation–π-cation interactions in proteins. Phys Chem Chem Phys 19:9849–9861
27. Vijay D, Sastry GN (2008) Exploring the size dependence of cyclic and acyclic π-systems on cation–π binding. Phys Chem Chem Phys 10:582–590
28. Mecozzi S, West AP, Dougherty DA (1996) Cation−π interactions in simple aromatics: electrostatics provide a predictive tool. J Am Chem Soc 118:2307–2308
29. Moreno DV, González SA, Reyes A (2010) Secondary hydrogen isotope effects on the structure and stability of cation–π complexes (Cation = Li^+, Na^+, K^+ and π = Acetylene, Ethylene, Benzene). J Phys Chem A 114:9231–9236
30. Yamada S, Misono T, Tsuzuki S (2004) Cation−π interactions of a thiocarbonyl group and a carbonyl group with a pyridinium nucleus. J Am Chem Soc 126:9862–9872
31. Gallivan JP, Dougherty DA (2000) A computational study of cation–π interactions vs salt bridges in aqueous media: implications for protein engineering. J Am Chem Soc 122:870–874
32. Marshall MS, Steele RP, Thanthiriwatte KS, Sherrill CD (2009) Potential energy curves for cation−π interactions: off-axis configurations are also attractive. J Phys Chem A 113:13628–13632
33. Mahadevi AS, Sastry GN (2016) Cooperativity in noncovalent interactions. Chem Rev 116:2775–2825
34. Quinonero D, Garau C, Rotger C, Frontera A, Ballester P, Costa A, Deya PM (2002) Anion-π interactions: do they exist? Angew Chem Int Ed 41:3389–3392
35. Alkorta I, Rozas I, Elguero J (2002) Interaction of anions with perfluoro aromatic compounds. J Am Chem Soc 124:8593–8598
36. Garau C, Quiñonero D, Frontera A, Ballester P, Costa A, Deyà PM (2003) Anion–π interactions: must the aromatic ring be electron deficient? New J Chem 27:211–214
37. Azizi A, Ebrahimi A (2019) The effects of anion approaching directions to the π-π^+ interaction. J Mol Liq 276:170–178
38. Reddy AS, Vijay DG, Sastry GM, Sastry GN (2006) From subtle to substantial: role of metal ions on π-π interactions. J Phys Chem B 110:2479–2481
39. Kumpf RA, Dougherty DA (1993) A mechanism for ion selectivity in potassium channels: computational studies of cation–π interactions. Science 261:1708–1710
40. Liu T, Zhu W, Gu J, Shen J, Luo X, Chen G, Puah CM, Silman I, Chen K, Sussman JL, Jiang H (2004) Additivity of cation–π interactions: an ab initio computational study on π-cation–π sandwich complexes. J Phys Chem A 108:9400–9405
41. Frontera A, Quiñonero D, Garau C, Costa A, Ballester P, Deyà PM (2006) Interplay between cation–π and hydrogen bonding interactions. J Phys Chem A 110:9307–9309
42. Yamada S (2020) Cation–π interactions in organic crystals. Coord Chem Rev 415:213301
43. Yamada S, Nojiri Y (2011) Water-assisted assembly of (E)-arylvinylpyridine hydrochlorides: effective substrates for solid-state [2+2] photodimerization. Chem Commun 47:9143–9145
44. Yamada S, Tokugawa Y (2009) Cation–π controlled solid-state photodimerization of 4-azachalcones. J Am Chem Soc 131:2098–2099

45. Escudero D, Frontera A, Quiñonero D, Deyà PM (2008) Interplay between cation–π and hydrogen bonding interactions. Chem Phys Lett 456:257–261
46. Rooman M, Liévin J, Buisine E, Wintjens R (2002) Cation–π/H-bond stair motifs at protein–DNA interfaces. J Mol Biol 319:67–76
47. Wu R, McMahon TB (2008) Investigation of cation–π interactions in biological systems. J Am Chem Soc 130:12554–12555
48. Lu Y, Liu Y, Li H, Zhu X, Liu H, Zhu W (2012) Mutual influence between halogen bonds and cation–π interactions: a theoretical study. ChemPhysChem 13:2154–2161
49. Pramanik A, Majumdar S, Das G (2010) Aryl azo imidazoles assisted assembly of anion/anion–water through salt formation. CrystEngComm 12:250–259
50. Yamada S (2018) Cation–π Interactions in organic synthesis. Chem Rev 118(23):11353–11432

Chapter 3
Biological Systems

3.1 Protein Structure

Polypeptide chains produced from amino acids by peptide bonds form secondary structures, which are regular local substructures such as α-helices and β-sheets (Fig. 3.1). These α-helices and β-sheets are folded to form a three-dimensional structure named as the tertiary structure. Some proteins are known to form quaternary structures consisting of two or more polypeptide chains or subunits, which function as a single functional unit.

These structures are stabilized by hydrogen bonds, electrostatic interactions, and π-π and CH-π interactions within and between peptide chains. In recent years, it has become clear that cation-π interactions significantly contribute to the formation of 3D structures of proteins, and their role in the biological processes has attracted considerable attention.

3.1.1 PDB Analysis of Protein Structures

Gallivan and Dougherty systematically surveyed the frequency of interactions between cations and π-planes in proteins from the PDB database to evaluate whether specific cation-π pairs are prioritized [1]. Using an originally developed program to detect cation-π interactions in protein crystal structures obtained from PDB, 2,994 significant cation-π interactions were detected out of a total of 230,504 amino acid residues (Table 3.1). Many of the detected cation-π interactions have very large energies, with about a quarter of them having electrostatic energy (E_{es}) greater than −4.0 kcal/mol.

S. Yamada, *The Cation-π Interaction*,
https://doi.org/10.1007/978-981-19-7335-2_3

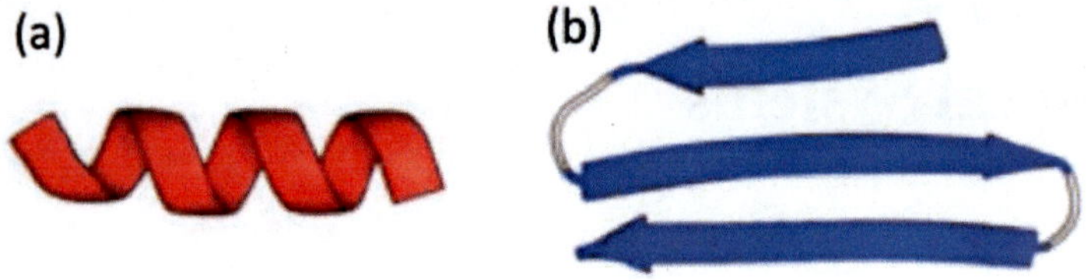

Fig. 3.1 Schematic representation of **a** α-helix and **b** β-sheet

Table 3.1 Frequency of cation-π interactions within proteins

Amino acid	Total number[a]	Interacting pairs[b]	Cation-π interactions[c]	Amino acid pair	Cation-π interactions[c]	Percent[d]
K	13,446	5,881	1,006	KF	285	14.5
R	10,919	8,149	1,988	KY	438	14.7
F	9,162	4,969	915	KW	283	30.2
Y	8,309	6,615	1,187	RF	630	21.0
W	3,412	2,446	892	RY	749	20.6
				RW	609	40.4

[a] The total number of times a particular amino acid appears in the data set of 593 proteins. [b] The number of times a particular amino acid occurs in an interacting pair. [c] The number of times a particular amino acid or pair of amino acids occurs in a cation-π interaction. [d] Percent of interacting pairs that are energetically significant cation-π interactions

Table 3.1 shows that Trp(W) is the aromatic residue most likely to be involved in cation-π interactions; 892 of 3412 are involved in cation-π interactions, corresponding to 26% of all Trps. This is in agreement with the theoretical prediction that Trp side chains are particularly well suited for cation-π interactions. They also find that Arg(R) is more likely than Lys(K) to be involved in energetically significant cation-π interactions. The most frequent combination is Arg/Trp(RW), with a frequency of 40%. There are 2,994 cation-π interactions on 230,504 residues, which means that there is an average of one energetically significant cation interaction for every 77 residues of the protein in the calculation.

3.1.2 *α-Helix*

The most common cation-π interactions are those between neighboring residues in the sequence, with 7.3% of interactions occurring between adjacent residues. The interactions between residues at the *i* and (*i*+4) positions are the second most common, suggesting that cation-π interactions often occur within the α-helix. Figure 3.2 shows the structure of the vaccinia virus protein VP39 (PDB ID code: 1V39), where Tyr83 and Arg79 can be seen in close proximity to each other. These results show the need to consider cationic interactions in the analysis of protein

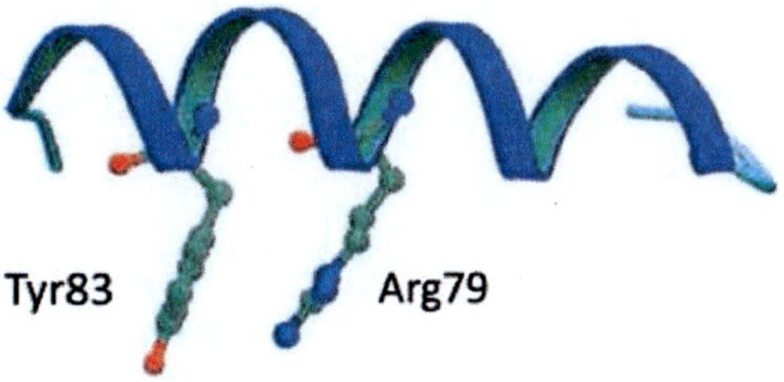

Fig. 3.2 An example of a strong cation-π interaction in an α-helix (Ees 4.2 kcal mol). The plot was created by using MOLSCRIPT and RASTER3D. Reprinted with permission from [1]. Copyright 1999 National Academy of Sciences

Table 3.2 Observed helicity and energetics for the peptides

Peptide	Sequence	$-[\theta]_{222} \times 10^{-3}$ (deg cm^2 dmol^{-1})	ϕH	$\Delta\Delta G$ (kcal mol^{-1})[a]
WR5	Ac-OOAAAAWAAAARAAAAOO-NH_2	11.1	0.347	0
WR4	Ac-OOAAAAAWAAARAAAAOO-NH_2	13.5	0.422	−0.4
RW5	Ac-OOAAAARAAAAWAAAAOO-NH_2	9.66	0.302	–
RW4	Ac-OOAAAARAAAWAAAAAOO-NH_2	8.68	0.271	–

[a] The free energies of side chain interactions were computed using an algorithm based on the Zimm-Bragg helix–coil transition model, with the nucleation parameter $\sigma = 0.004$ and values of the helix propensities from amino acids Ala, Trp, Arg, Glu, Phe, Tyr, and Orn

structures, along with more general hydrogen bonding, salt bridges, and hydrophobic effects.

Shi et al. performed CD measurements on the alanine-based peptide WR4 and its control compound WR5 to investigate the effect of the interaction between the Trp and Arg residues on the stability of the helix [2]. The Trp and Arg side chains of WR4 are located at (*i*, *i*+4) positions on the surface of the helix, and the Trp and Arg side chains of WR5 are located at (*i*, *i*+5) positions that prevent the interaction between them. In these two peptides, the Arg residue is in the same position but the position of Trp is different. In order to further investigate the effect of orientation on the interaction, RW4 / RW5 was designed (Table 3.2).

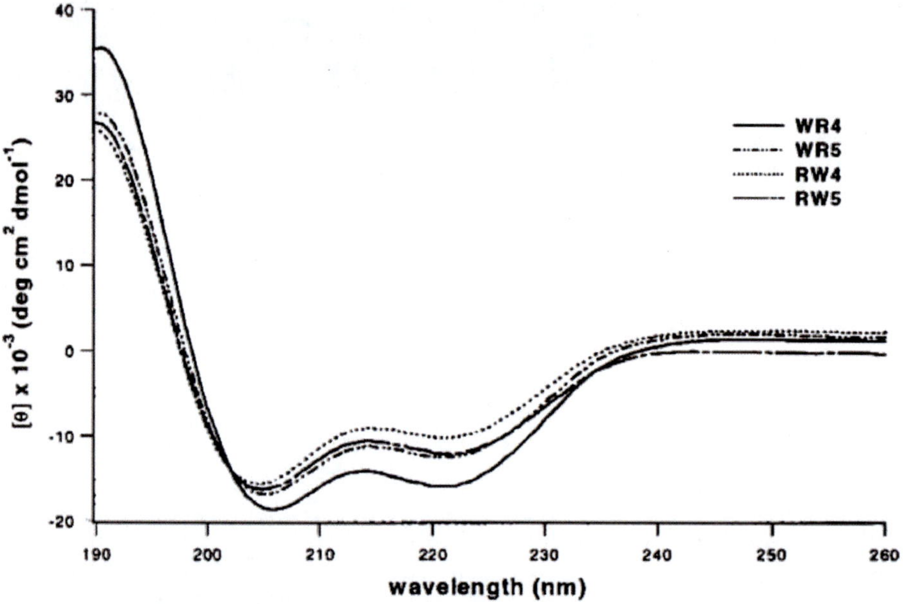

Fig. 3.3 Circular dichroism spectra of W/R pe tides in 10 mM phosphate buffer pH 7 at 4 °C. Reprinted with permission from [2]. Copyright 2002 American Chemical Society

From the CD spectra shown in Fig. 3.3, all the peptides show minima at 208 nm and 222 nm, clearly showing that they have helical structures. A comparison of the stabilization energies from the CD spectra shows that the WR4 helix is more stable than the control peptide WR5 by 0.4 kcal/mol, indicating that the Trp→Arg (*i*, *i*+4) interaction is effective in stabilizing the α- helix. On the other hand, when the orientation of the two residues is reversed, no increase in the Arg→Trp (*i*, *i*+4) free energy is detected. This suggests that the orientation of the amino acid sequence is important for the interaction and stabilizes the helix only in the N→C direction.

Besides the stabilization of α-helices by the Trp/Arg (Trp—guanidinium) interaction described above, it has been shown that ammonium ions also contribute to the stabilization of α-helices. Tsou et al. investigated the magnitude of the cation-π interactions between phenylalanine (Phe), lysine (Lys), ornithine (Orn), and diaminobutanoic acid (Dab) in α-helical peptides, F8X12 and F7X12, by CD spectroscopy [3]. The results are shown in Table 3.3. Comparing F8K12 (*i*, *i*+4) with control F7K12 (*i*, *i*+5), there was no difference in the CD spectra, indicating that there is almost no stabilizing effect of the cation-π interaction in Phe/Lys. On the other hand, the spectrum of F8O12 with a different side chain length showed that the peptide is stabilized by −0.4 kcal/mol.

Table 3.3 Mean residue ellipticities, helicities, and side chain-side chain interaction energies of peptides

(a) **F8X12**: Ac-AAKAAKA-**F**AAA**X**AK-AAKK-GGY-NH_2

(b) **F7X12**: Ac-AKAAKA**F**-AAAA**X**AK-AAKK-GGY-NH_2

X = K(Lys), O(Orn) or Dab

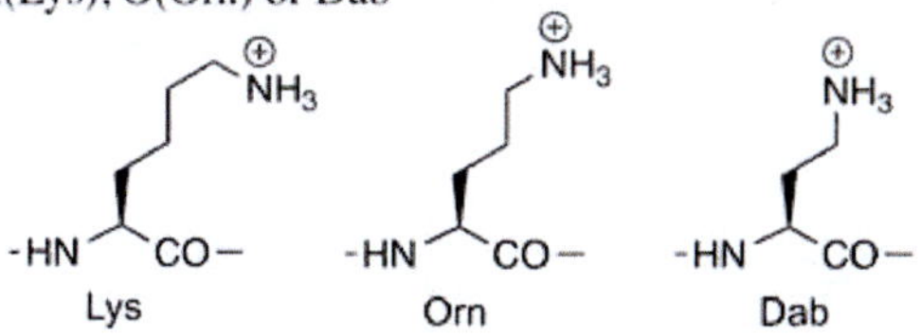

Peptide	$[\theta]_{222}$ (deg cm^2 dmol^{-1})	%helix (±3%)	$\Delta\Delta G^a$ (kcal/mol)
F8K12	−18.800	52	0
F7K12	−17.900	49	0
F8O12	−15.900	43	−0.4
F7O12	−11.500	32	0
F8Dab12	−11.600	34	ND
F7Dab12	−9.900	29	ND

[a] $\Delta G°$ is the interaction energy between Phe and Lys or Orn. It was determined with AGADIR, by allowing the interaction energy to vary until the calculated helicity fits the observed helicity

Furthermore, strong NOEs observed in Phe and Orn by NMR measurements at 25 °C confirm the presence of the interaction between them. On the other hand, NOEs were also observed between Phe and Dab and Phe and Lys at 5 °C, indicating a weak interaction between them. These results show that even a single ammonium-π interaction can provide significant stability to a protein structure.

3.1.3 *β-Hairpin*

Tatko and Waters investigated the contribution of cation-π interactions to the stability and specificity of hairpin peptides [4]. For model hairpin peptides Ac-RX_1VEVNGOX_2ILQ-NH_2 (peptide A) (Fig. 3.4), the effect of cation-π interactions between Phe or Trp and Lys or Arg at the diagonal position was evaluated. When the substituents X_1 and X_2 of peptide A were Phe/Lys (FK), Trp/Lys (WK), and Val/Lys (VK), the Ha chemical shifts of peptide A and the difference in the chemical shifts from the control cyclic peptide (δΔ values) were examined (Fig. 3.5a). The largest values were obtained for Trp compared to Val and Phe. This suggested that the best β-hairpin structure was taken at Trp/Lys(WK).

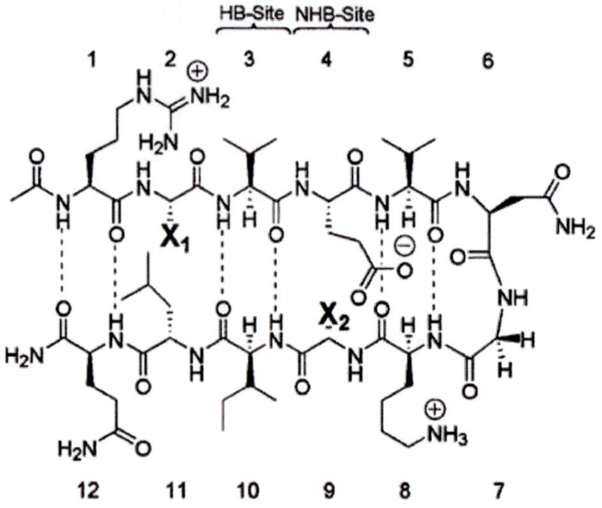

Fig. 3.4 Model peptide system Ac-RX_1VEVNGOX_2ILQ-NH_2 (peptide A) used to probe the cation-π diagonal interaction. Peptides are referred to by the diagonal residues in the text (X_1, X_2). HB-site indicates hydrogen-bonded site; NHB-site, non-hydrogen-bonded site. From [4] Copyright © 2003 by John Wiley Sons, Inc. Reprinted by permission of John Wiley & Sons, Inc

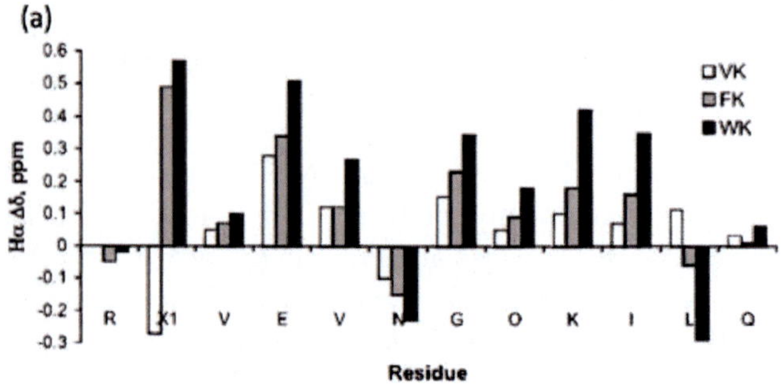

(b)

Peptide	ΔΔ*G* (kcal/mol)
FK	-0.20
FR	-0.29
WK	-0.35
WR	-0.48

Fig. 3.5 **a** Difference in β-hairpin and random coil shifts. Asn is shifted upfield, as is expected for the turn residue. The Gly Hα is the difference in chemical shift of the diastereotopic Hαs. Residues Arg1 and Gln12 are frayed and show little shifting relative to the random coil values. **b** The diagonal interaction energies as determined from double-mutant cycles. From [4] Copyright © 2003 by John Wiley Sons, Inc

Figure 3.5b shows the stabilization energies calculated by the double-mutant cycle method. The stabilization energies due to the interaction between substituents X_1 and X_2 at the diagonal position ranged from −0.20 to −0.48 kcal/mol. As cationic residue R showed higher stabilization than K, aromatic amino acid residue Trp was found to be more effective in stabilizing than Phe.

Methylation of lysine in histone proteins is a post-translational modification during the process of chromosome condensation. This phenomenon has been linked to epigenetic silencing, and it has been suggested that these methylated bodies act as mediators of protein–protein interactions. In particular, methylated lysines in histone tails have been shown to trigger protein–protein interactions with several proteins, including chromodomains, in which $Lys(Me)_n^+$ (n = 1−3) binds within an aromatic pocket composed of three Trp and Tyr side chains. Hughes and Waters constructed a model system for lysine methylation in histone proteins and investigated the effect of lysine methylation on the magnitude of the cation-π interaction and the stability of the designed β-hairpin [5]. In these studies, they created the most stable β-hairpin reported to date.

Figure 3.6 shows the δΔ values, which are differences in the ^{1}H NMR chemical shifts of the Ha of peptides **1** and **2** with that of control peptides **3**–**7**. These led to the extent of folding of **1** and **2** to be 78% and 92%, respectively. This corresponds to ΔG values of −0.75 and −1.45 kcal/mol. From these results, it is estimated that methylation improves the overall stability of the hairpin by about −0.7 kcal/mol.

In addition, the estimated interaction for Lys···Trp and $Lys(Me)_3$···Trp by double-mutant cycle is −0.4 and −1.0 kcal/mol, respectively. The reason for the increase in hairpin stability upon methylation can be explained by the increase in the magnitude of the cation-π interaction due to effective overlap of Trp with $Lys(Me)_3$ than that with Lys as shown in Fig. 3.7.

Riemen and Waters designed very well folded hairpin peptides that mimic binding motifs found in natural protein–protein interactions [6]. The hairpin peptides (Fig. 3.8a) have two tryptophan residues and form an aromatic pocket, surrounding the lysine side chain by cation-π interactions as observed in trimethylated lysine receptor proteins.

(a) Ac-Arg-Trp-Val-Glu-Val-Asn-Gly-Orn-**X**-Ile-Leu-Gln-NH_2
1: X = Lys; **2**: X = $Lys(Me)_3^{\oplus}$

(b) Ac-<u>Cys</u>-Arg-Trp-Val-Glu-Val-Asn-Gly-Orn-**X**-Ile-Leu-Gln-<u>Cys</u>-NH_2
3: X = Lys; **4**: X = $Lys(Me)_3^{\oplus}$

(c) **5** Ac-Asn-Gly-Orn-Lys-Ile-Leu-Gln-NH_2
6 Ac-Asn-Gly-Orn-$Lys(Me)_3^{\oplus}$-Ile-Leu-Gln-NH_2
7 Ac-Arg-Trp-Val-Glu-Val-Asn-Gly-NH_2

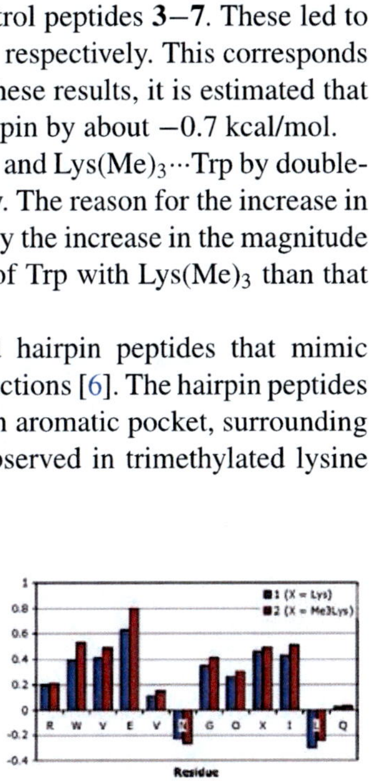

Fig. 3.6 **a** β-hairpin peptides **1** and **2**. **b** Control peptides for the fully folded state cyclized with a Cys-Cys disulfide bond between the underlined residues. **c** Control peptides for the unfolded state. HR shifts of peptides **1** and **2**. The Gly bars reflect the HR separation in the hairpin. Reprinted with permission from [5]. Copyright 2005 American Chemical Society

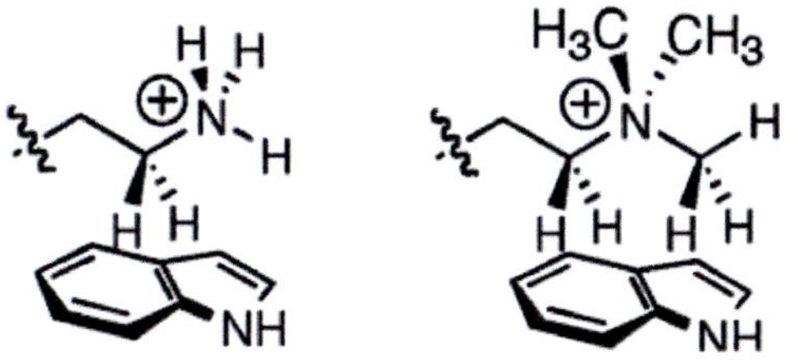

Fig. 3.7 Possible interaction geometries for Trp···Lys and Trp···$Lys(Me)_3^+$ interactions. Reprinted with permission from [5]. Copyright 2005 American Chemical Society

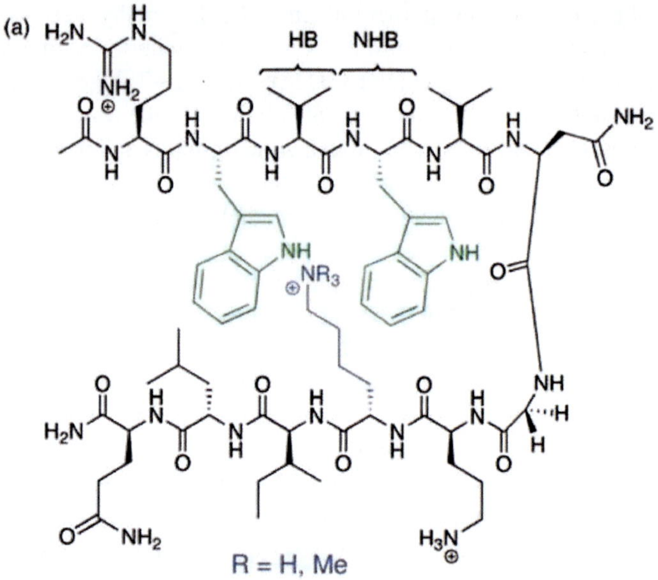

(b) Ac-Cys-Arg-Trp-Val-Trp-Val-Asn-Gly-Orn-KMe$_n$-Ile-Leu-Gln-Cys-NH$_2$

(c) Ac-Arg-Trp-Val-Trp-Val-Asn-Gly-NH$_2$

Ac-Asn-Gly-Orn-KMe$_n$-Ile-Leu-Gln-NH$_2$

n = 0 - 3

Fig. 3.8 **a** Trp pocket series of peptides containing Trp at positions 2 and 4 and Lys, mono-, di-, or trimethyl-Lys at position 9 (R=H or Me, accordingly). HB=hydrogen-bonded site; NHB=non-hydrogen-bonded site. **b** Sequence of control peptides for the fully folded state cyclized with a Cys-Cys disulfide bond between the underlined residues. **c** Sequence of control peptides for the unfolded state. Reprinted with permission from [6]. Copyright 2009 American Chemical Society

Thermal analysis of methylated hairpin peptides revealed that thermal stability increased with increasing degree of methylation (Fig. 3.18a). The stability of the Trp pocket peptides is comparable to that of other highly stabilized hairpins such as trpzip4, HP5W4, and WKMe3 by a variety of stabilizing interactions (Table 3.4). The unmethylated Trp K pocket is 78% folded at 73 °C which is as stable as WKMe3. On the other hand, Trp KMe3 pocket hairpin peptide is more stable than the others and is 96% folded at 72 °C. Thus, a very stable β-hairpin structure can be generated by combining the cation-π interaction with a β-sheet structure containing Trp residues.

3.1.4 Miniature Proteins

The "WSXWS" motif is known to be involved in fibronectin type III (Fn3) domain containing proteins. This motif has been the focus of extensive biological studies, as it is found in a wide array of extracellular receptor (Table 3.5). This motif contains β-strand:loop:PPII-helix topology, and the cation-π interactions between the arginine

Table 3.4 Comparison of highly stable β-Hairpins

Peptide	Sequence	Fraction folded	
		At 25 °C	At 72–75 °C
trpzip4	GEWTWDDATKTWTWTE	92%	40%
HP5W4	KKWTWNPATGKWTWQE	>96%	60%
WKMe3	RWVEVNGOKMe3ILQ	92%	72%
Trp K pocket	RWVWVNGOKILQ	99% (93%)[a]	78% (65%)[a]
Trp KMe3 pocket	RWVWVNGOKMe3ILQ	99% (95%)[a]	96% (80%)[a]

[a] The value in parentheses comes from the average HR splitting of the residues in the hydrogen-bonded sites

residue of β-strand and the tryptophan residue of PPII-helix stabilize a relatively rare secondary structure, the left-handed polyproline type-II helix (PPII) (Fig. 3.9). As an example of a peptide having such sequence, the X-ray crystal structure of IL3RB (the common name for P32927) was shown in Fig. 3.10. This clearly shows the existence of a β-strand:loop:PPII-helix topology with close contact of the β-strand and the PPII-helix and the cation-π interaction network is formed between the arginine and tryptophan residues of the WSXWS motif.

Table 3.5 Uniprot IDs and multiple sequence alignment of motifs. WSXWS motifs found in X-ray crystal structures

Uniprot ID	Multiple sequence alignment
P40189	**R**I**R**CMKEDGKG....Y**WS**D**WS**E
P14784	QV**R**V**K**PLQGEFT...T**WS**P**WS**Q
P31785	**R**V**R**S**R**FNPLCGSAQ.H**WS**E**WS**H
P78552	**R**I**R**V**K**TNKLCYEDDKL**WS**N**WS**Q
P40223	QM**R**CIRSSLPG....F**WS**P**WS**P
P19235	AV**R**A**R**MAEPSFGG..F**WS**E**WS**E
Q99062	QI**R**CIRWPLPG....H**WS**D**WS**P
P42702	**R**I**R**CSTETFW.....K**WS**K**WS**N
P05710	QT**R**C**K**PDHG......Y**WS**R**WS**Q
Q14627	VV**R**S**K**VNIYCSDDG.I**WS**E**WS**D
P08887	QL**R**AQEEFGQG....E**WS**E**WS**P
Q01344	QV**R**AAVSSMCREAG.L**WS**E**WS**Q
Q9HBE5	QV**R**AGPMPGSSYQG.T**WS**E**WS**D
P48357	QV**R**C**K**RLDGLG....Y**WS**N**WS**N
P32927[a]	**R**V**R**V**R**TSRTGYNG..I**WS**E**WS**E

[a] The common name for P32927 is IL3RB

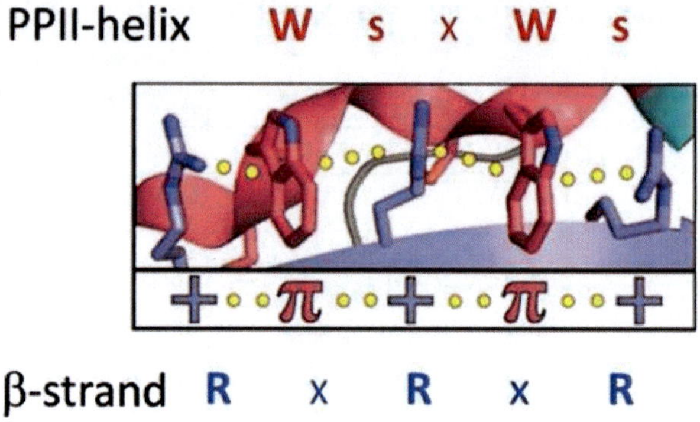

Fig. 3.9 Cation-π interaction network stabilizes "WSXWS motif." The dotted yellow lines highlight the connectivity of the cation-π interaction network formed between arginine (blue) and tryptophan (purple) side chains. Reprinted with permission from [7]. Copyright 2016 American Chemical Society

Fig. 3.10 X-ray crystal structure of IL3RB (the common name for P32927)

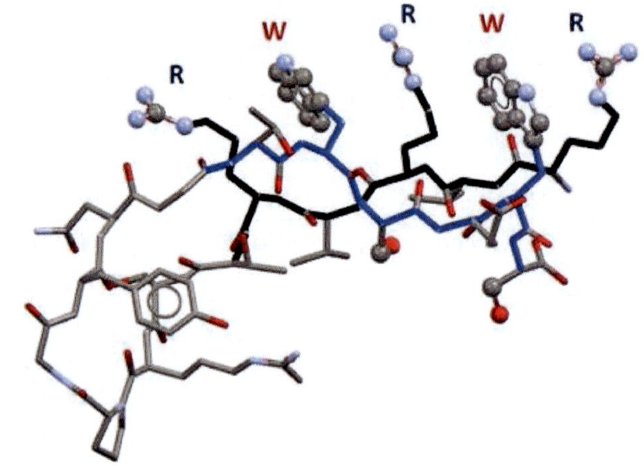

Craven et al. designed 19-residue miniature protein, RVRVRTSRpGYNGTWSEWS, containing the WSXWS motif to establish thermostable tertiary structures [7]. The CD spectra showed that it contained secondary and tertiary structures, with a stabilizing energy ΔG° at folding of about −6.7 kcal/mol. NOESY measurements by NMR suggest that arginine and tryptophan residues are rigidly assembled and have a highly organized 3D structure. Since the WSXWS motif is also found extensively in extracellular receptors and is a focus of biological research, continued study of the structure of such designed miniature proteins could contribute to the development of the next generation of peptide therapeutics and biochemical tools.

3.1.5 pH Dependence on Protein Conformation

It is known that some proteins change their structure and activity depending on pH. In this conformational change, the cation-π interaction makes a significant contribution to the resultant conformational stabilization. For example, Staphylococcus aureus secretes phosphatidylinositol-specific phospholipase C (PI-PLC), which is more active at acidic pH values than other enzymes. The crystal structures of S. aureus PI-PLC at pH 4.6 and pH 7.5 are shown in Fig. 3.11a, b, respectively. Furthermore, the crystal structure of the enzyme with H258 replaced by Y258 at pH 4.6 is shown in Fig. 3.11c [8].

Under weakly basic conditions (pH 7.5), the conformation of PI-PLC (Fig. 3.11b) is very close to those of other bacterial PI-PLCs. In contrast, when crystallized under acidic conditions (pH 4.6), the movable loop at the edge of the αβ-barrel rim near the active site shifts by about 10 Å (Fig. 3.11a). The crystal structure of the mutant protein obtained at pH 4.6 does not show a large loop shift (Fig. 3.11c) and is similar to the crystal structure at pH 7.5 (Fig. 3.11b). Figure 3.11d, which is the enlarged view of Fig. 3.11a, clearly shows that both residues His258 and Phe249 approach and interact with each other, strongly suggesting that the shape change of the PI-PLC loop at acidic pH is a result of intramolecular cation-π interaction between His258 and Phe249.

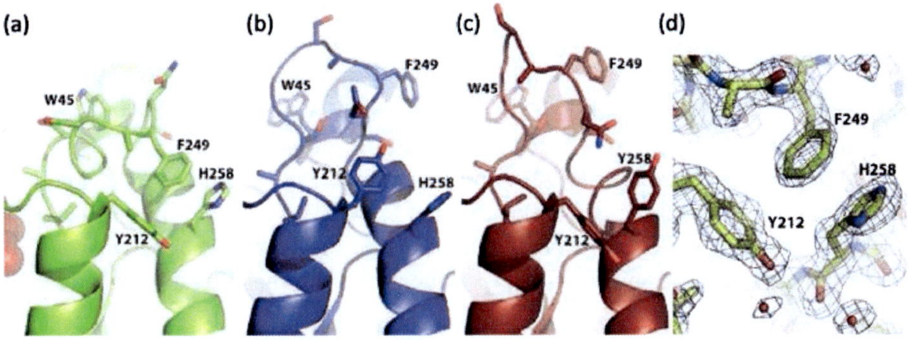

Fig. 3.11 A close-up of the S. aureus PI-PLC mobile rim loop conformation near helices F and G in the wild-type enzyme at **a** pH 4.6 and **b** pH 7.5, compared to the same region in the H258Y structure at pH 4.6 **c**. The proposed Phe249/His258 π-cation interaction and relevant side chains are indicated in each panel. **d** View of the π-cation interaction residues. Reprinted with permission from [8]. Copyright 2012 American Chemical Society

3.2 Protein–Protein Association

As described in Sect. 3.1, cation-π interactions contribute to stabilization of the secondary and tertiary structures of proteins. This section introduces that the generation or dissociation of quaternary structures involves cation-π interactions.

3.2.1 PDB Analysis of Protein–Protein Interfaces

In order to investigate protein–protein interactions, a protein database has been surveyed and clarified that about half of the protein complexes and one-third of the homodimers contain at least one intermolecular cation-π pair as shown in Table 3.6 [9]. Among the amino acid residues, arginine residues were found to be particularly abundant at the protein–protein interface, in which cation-π interactions with aromatic amino acid residue are involved. The frequency of the interaction with tyrosine was found to be the highest among the aromatic amino acid residues, and the calculated electrostatic interaction was 3 kcal/mol on average. In addition, the cationic part of the cation-π pair participates in intermolecular hydrogen bonding with high frequency.

Figure 3.12a shows a crystal structure of homodimer of farnesyl diphosphate synthase as a typical example, in which the symmetric cation-π interaction was observed between R155 and Y179 at the homodimer interface. Figure 3.12b shows an example of a complex of p53 protein bound to ankyrin and SH3 of 53PB2. In this interface, the arginine residue R248 interacts with W498 and Y509. In addition, it forms an ion pair with Asp475, to stabilizing the complex. Figure 3.12c, d show the Ran-importin beta and Ran-RCC1 complexes, respectively, where Arg/Tyr cation-π interactions are observed at R110/Y156 and R106/Y323. Furthermore, each guanidinium group forms a more robust complex due to salt bridge formation with the acidic groups of D160 and D384. Thus, as mentioned in Chap. 2, cation-π interactions

Table 3.6 Classes of protein–protein interfaces analyzed for cation–π interactions

Class	Sample size	Cation-π interactions	
		Interface[a]	Total[b]
Crystal-packing	116	14	23
Protein complex	81	40	68
Homodimers	120	42	141
Heterocomplexes	417	218	482
Total	734	314	714

[a] The number of interfaces in which at least one cation-π pair was detected. [b] The total number of cation-π pairs detected in each data set including the duplication arising from twofold symmetry

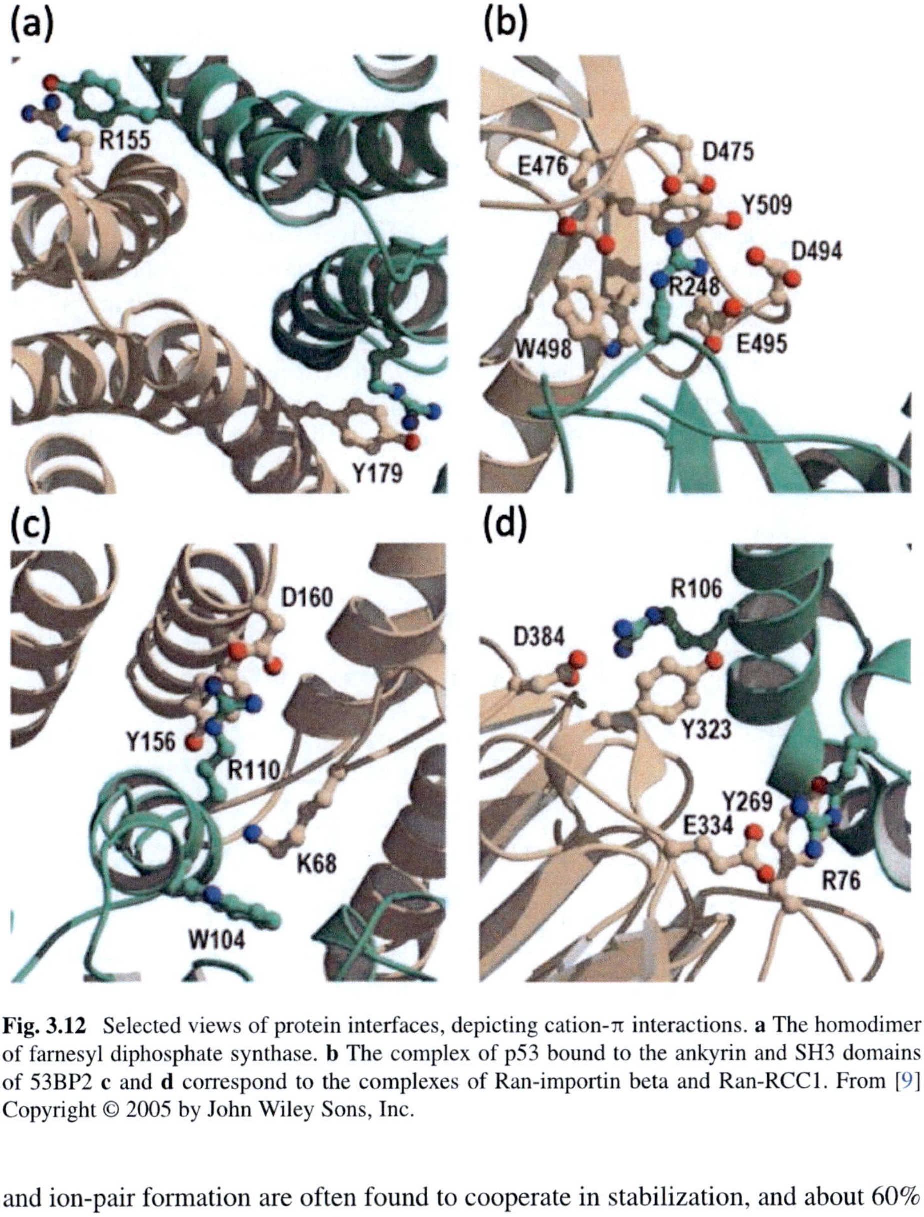

Fig. 3.12 Selected views of protein interfaces, depicting cation-π interactions. **a** The homodimer of farnesyl diphosphate synthase. **b** The complex of p53 bound to the ankyrin and SH3 domains of 53BP2 **c** and **d** correspond to the complexes of Ran-importin beta and Ran-RCC1. From [9] Copyright © 2005 by John Wiley Sons, Inc.

and ion-pair formation are often found to cooperate in stabilization, and about 60% of the protein complex interfaces involving cation-π interactions form salt bridges.

3.2.2 *Helix–Helix Association*

Johnson et al. found that two helix transmembrane segments rich in Ala and Ile (isoleucine) tend to form SDS-resistant dimers when Trp residues are properly positioned [10]. The combination of Ala and Ile residues is denoted as AI, and the two reverse-parallel helix transmembrane segments are best joined. The authors systematically investigated the influence of cation-π interactions in the hydrophobic segment AIAIAIIAZAXAIIAIAIAI on tertiary structure formation in E. coli membranes [11]. Here, aromatic residues Z=W, Y, or F, and cationic residues X=H, R, or K. All possible combinations of pairings were evaluated for dimerization of the transmembrane (TM) domain by TOXCAT oligomerization assay.

The sequencing of the various segments used is shown in Fig. 3.13. Figure 3.14 shows the results of TOXCAT evaluation of wild type (WT), glycophorin A (GpA), AI WT (WAI), and various AI mutants. From Fig. 3.14a, it is clear that dimerization of WAA, YAA, and FAA having aromatic residues is enhanced compared to the control (AAA). When A of them is replaced by K, which are indicated as WAK, YAK, and FAK, the strength of their oligomerizations was fourfold stronger than that of GpA WT, suggesting significant enhancement of cation-π interactions with Trp, Tyr, or Phe (Fig. 3.14c).

In addition, it is clear from the comparison of Fig. 3.14b and d that dimerization is promoted when both the basic residues (histidine, arginine, and lysine) and the aromatic residue (tryptophan) are present in the oligomers. These results

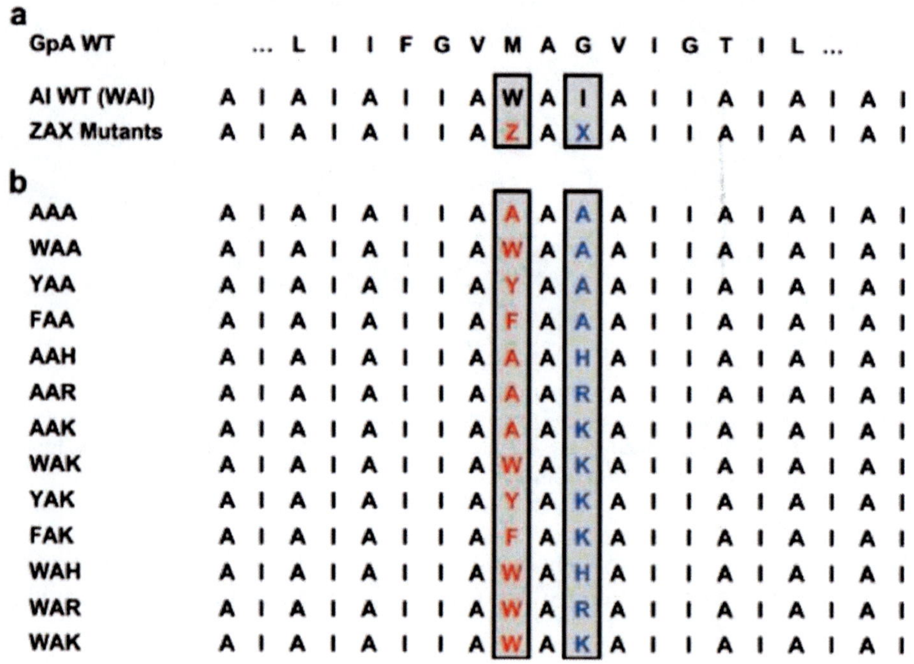

Fig. 3.13 Sequence alignment of AI aromatic and basic residue constructs. **a** WT GpA and WT AI (WAI) sequences along with the design of ZAX mutants. **b** Sequences of ZAX mutants are shown. Gray boxes represent the mutation positions, where aromatic residue positions are highlighted in red (Z = A, W, Y, or F) and the basic residue positions are highlighted in blue (X = A, H, K, or R). Reprinted with permission from [11]. Copyright 2017 American Chemical Society

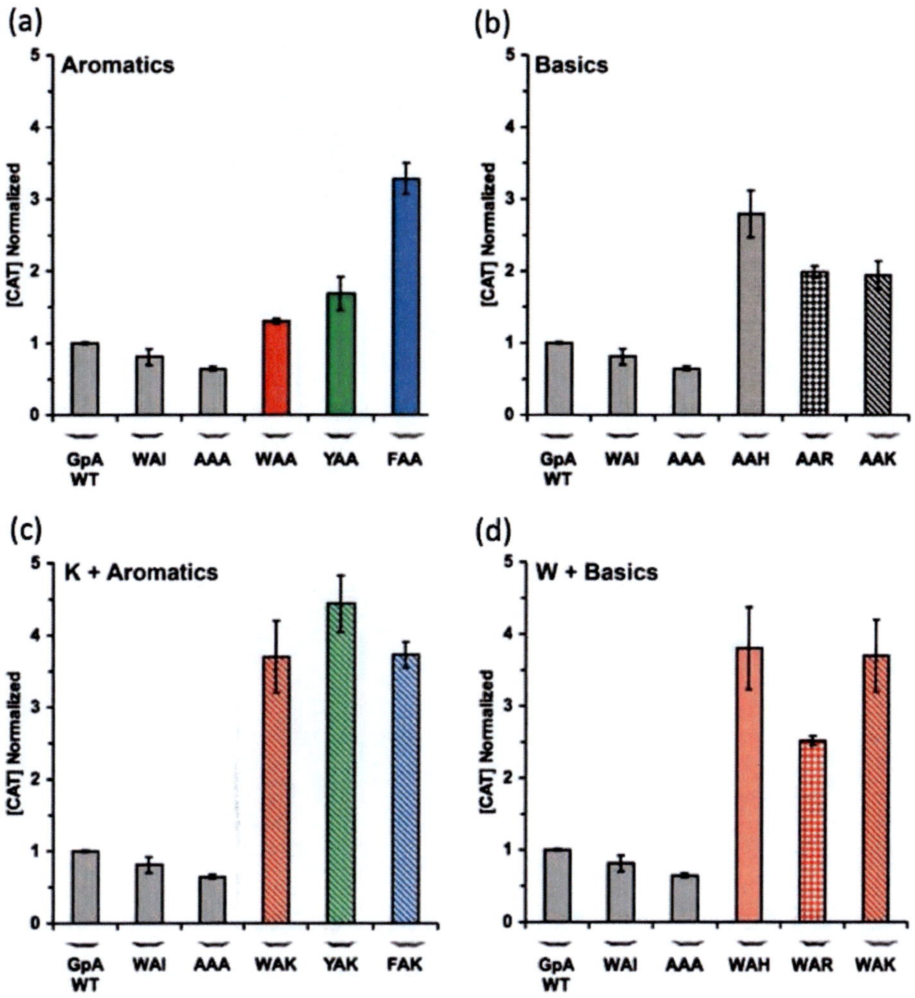

Fig. 3.14 TOXCAT assay of WT GpA and the AI aromatic and basic residue-substituted series. CAT expression is normalized to WT GpA and is shown for the following series: **a** aromatics; **b** basics; **c** Lys and aromatics; and **d** Trp and basics. Aromatic residue substitutions are colored in red (Trp), green (Tyr), or blue (Phe). Reprinted with permission from [11]. Copyright 2007 American Chemical Society

indicate that the TM–TM association of AI is promoted by the cation–π interaction. These systematic evaluations of hydrophobic helical oligomer formation in E. coli membranes showed that cation–π interactions play an important role in the hydrophobic membrane environment.

Figure 3.15 shows energy-minimized molecular model of a parallel dimer of the WAK construct. The Trp^{10} residue is highlighted in green, and the Lys^{12} residue is highlighted in red. These structures clarified that the Trp^{10} residues in one helix and Lys^{12} residues in the other helix are close together, indicating the contribution of cation-π interactions between them.

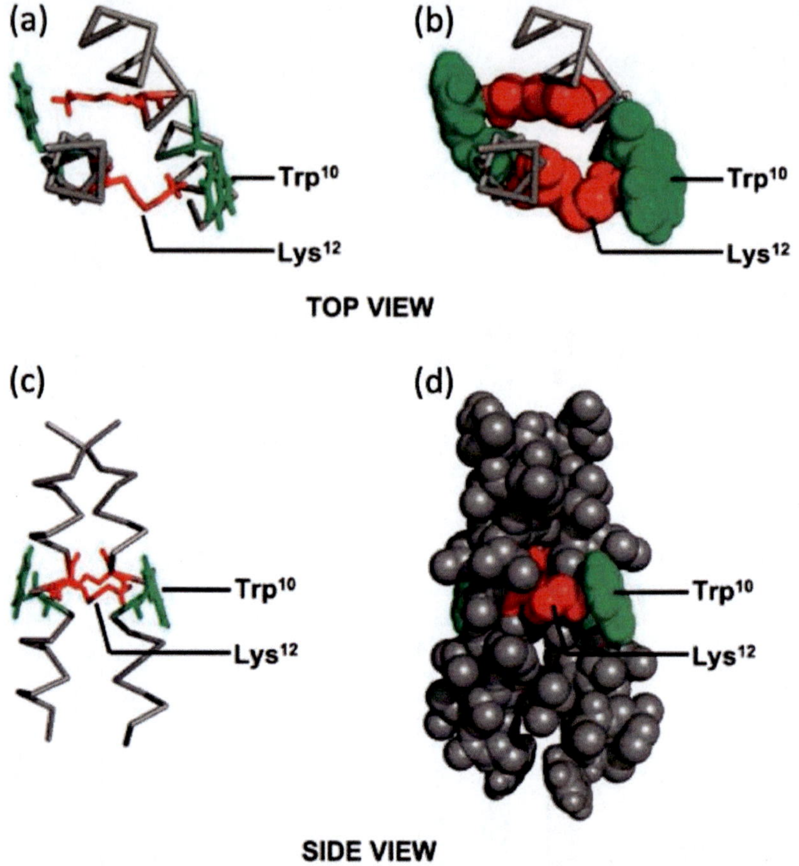

Fig. 3.15 Energy-minimized molecular model of a parallel dimer of the WAK construct. Backbone models are shown in **a** and **c**; space-filling models are shown in **b** and **d**. Both top views (panels a and b) and side views (panels c and d) are shown. Reprinted with permission from [11]. Copyright 2007 American Chemical Society

3.2.3 Self-Assembly of Collagen-Mimetic Peptides

Collagen is the most abundant protein in the human body, accounting for 25−35% of the total body protein content. Natural collagen is characterized by the repetition of X−Y-Gly units, with a high frequency of proline (Pro) and 4-hydroxyproline (Hyp,O) in the X and Y positions, respectively, with Pro-Hyp-Gly (POG) being the most common unit found in collagen. As most natural collagen is a heterotrimer consisting of two (AAB) or three (ABC) different peptide chains, the construction of heterotrimers that mimic natural collagen has received considerable attention. Chen et al. have investigated the construction of heterotrimers by utilizing intermolecular cation-π interactions [12].

First, they chose Arg-Phe as the cation-aromatic pair and designed the collagen-related peptide RG(POG)$_{10}$F, in which Arg is attached to the N-terminus and

Phe to the C-terminus of the peptide (Fig. 3.16). The structural properties of this peptide were investigated by circular dichroism (CD) measurements using $(POG)_{10}$, $RG(POG)_{10}$, and $G(POG)_{10}F$ as control peptides, and the Tm value of $RG(POG)_{10}F$ was found to be 67.4 °C, which is higher than that of all the control peptides ($(POG)_{10}$: 59.6, $(POG)_{10}$: 59.6, $RG(POG)_{10}$: 65.9, $G(POG)_{10}F$: 59.6). This indicates that $RG(POG)_{10}F$ forms a more stable collagen triple helix than do the control peptides.

Furthermore, the fibril growth of $RG(POG)_{10}F$ was monitored by dynamic light scattering (DLS) spectroscopy. As shown in Fig. 3.17, at a concentration of

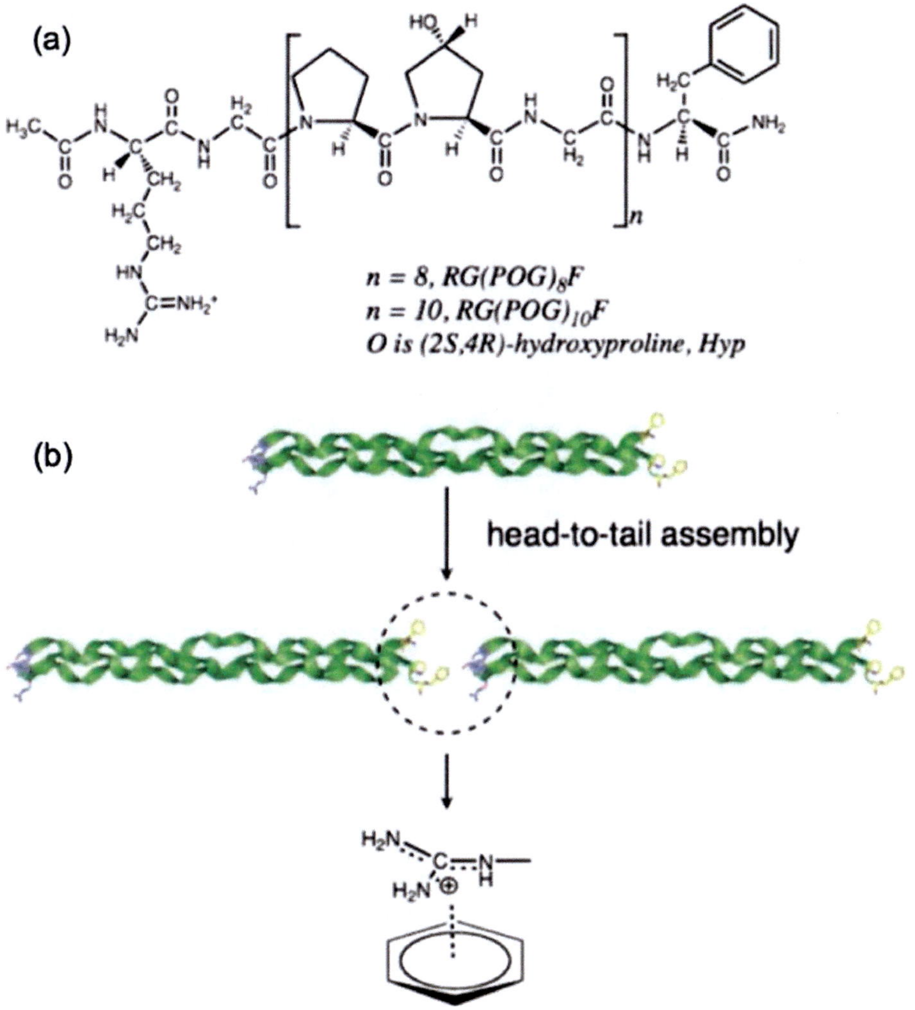

Fig. 3.16 Self-assembly of collagen-related peptides. **a** Designed peptide sequences of $RG(POG)_{10}F$ and $RG(POG)_8F$. **b** Illustration of the head-to-tail assembly of collagen triple helices via cation-π interactions. Arg residues are colored blue and Phe residues yellow. Reprinted with permission from [12]. Copyright 2011 American Chemical Society

1.0 mg/mL, its hydrodynamic diameter *d* increased from ~350 to ~1400 nm with an extension of its incubation time from 0 to 24 h (Fig. 3.17a). When the concentration was increased to 3.0 mg/mL, the size increased regardless of incubation time (Fig. 3.17b). On the other hand, DLS measurements of $(POG)_{10}$, $G(POG)_{10}F$, and $RG(POG)_{10}$ showed that their hydrodynamic diameters were all less than 150 nm upon dissolution in solution, indicating that they self-assemble much more slowly than $RG(POG)_{10}F$. These results indicate that the incorporation of cationic or aromatic residues at the ends of collagen-related peptides can promote the formation of triple-helical structures and fibrils in a head-to-tail fashion by cation-π interactions between Arg and Phe, as shown in Fig. 3.16.

Next, they designed and synthesized three types of collagen-mimetic peptides R, F, and Y (Fig. 3.18) incorporating Arg as a cationic group in Y and Phe and Tyr as aromatic groups in X of the constituent X–Y-Gly unit of the collagen and examined whether cation-π interactions could induce folding of the heterotrimers [13]. Detailed

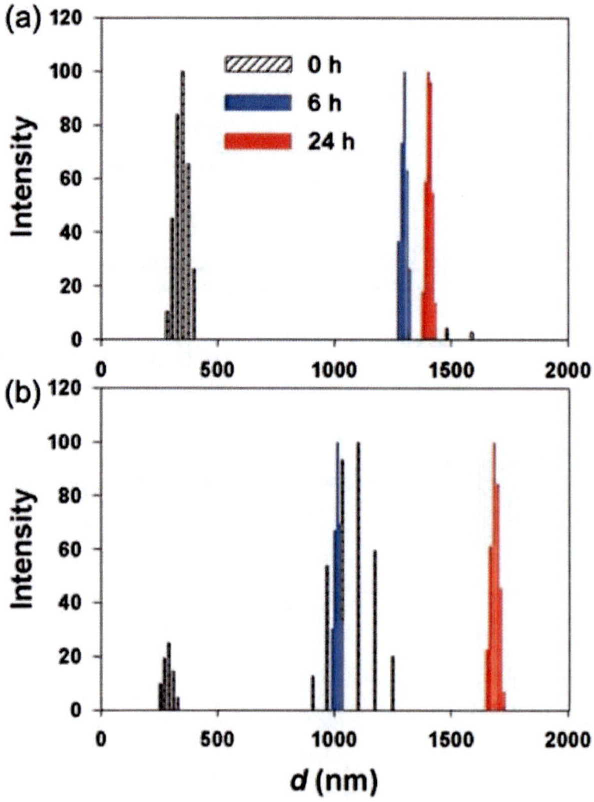

Fig. 3.17 Dynamic light scattering measurements for $RG(POG)_{10}F$ at concentrations of **a** 1.0 and **b** 3.0 mg/mL for different incubation times (0, 6, and 24 h) in 20 mM phosphate buffer (pH 7.0). The measurements were taken at 25 °C (0 h incubation) or 65 °C (6 and 24 h incubation). The X-axis d represents the hydrodynamic diameter. Reprinted with permission from [12]. Copyright 2011 American Chemical Society

(a)

R: Ac-(Pro-Hyp-Gly-Pro-Arg-Gly)$_4$(Pro-Hyp-Gly)-NH_2
F: Ac-(Pro-Hyp-Gly-Phe-Hyp-Gly)$_4$(Pro-Hyp-Gly)-NH_2
Y: Ac-(Pro-Hyp-Gly-Tyr-Hyp-Gly)$_4$(Pro-Hyp-Gly)-NH_2

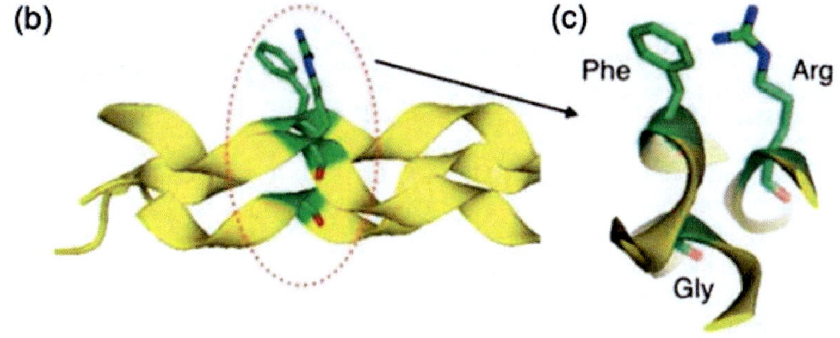

Fig. 3.18 **a** Sequences of the peptides R, F, and Y, where Hyp is 4*R*-hydroxyproline. All the peptides have an acetylated N-terminus and an amidated C-terminus. **b** Illustration of the interchain cation-π interaction in an R-F heterotrimer. **c** The relative position of Phe, Arg, and Gly viewed along the axis from the N-terminus. The diagrams were generated using the PDB entry 1CAG. Reprinted with permission from [13]. Copyright 2016 American Chemical Society

thermodynamic analysis of the synthesized AAB-type collagen heterotrimers by DSC showed that the cation-π interaction contributes to the folding of the AAB-type heterotrimers. Furthermore, kinetic measurements showed that the introduced cation-π interaction accelerates the folding of the heterotrimer (Fig. 3.18). These analyses provide new and valuable insights into the folding of collagen heterotrimers.

Chiang et al. further designed and synthesized cationic collagen-mimetic peptides (CMPs) with multiple Arg or Lys residues at the C-terminus (CR and CK) and aromatic CMPs with multiple Phe residues at the C-terminus (CF) (Fig. 3.19) [14]. The CD and NMR spectra showed that AAB-type heterotrimers were formed in the CR-CF and CK-CF mixtures, suggesting that the cation-π interaction between the C-terminal cation and aromatic residues acts as a nucleation force and greatly facilitates the folding of the heterotrimers. It is noteworthy that only one major heterotrimer was found in each mixture. On the other hand, in the CK-CF mixture, even when the ratio of CF was increased, only the heterotrimer consisting of two CK chains and one CF chain was formed in solution, suggesting that the heterotrimer consisting of one CK chain and two CF chains was very unstable. These results can be applied to the preparation of various collagen-related biomaterials.

3.2.4 *Dissociation of UVR8 Homodimer*

The Arabidopsis protein UVR8 is a UVB photoreceptor. When UVR8 is irradiated, it immediately dissociates from a homodimer to two monomers, which triggers

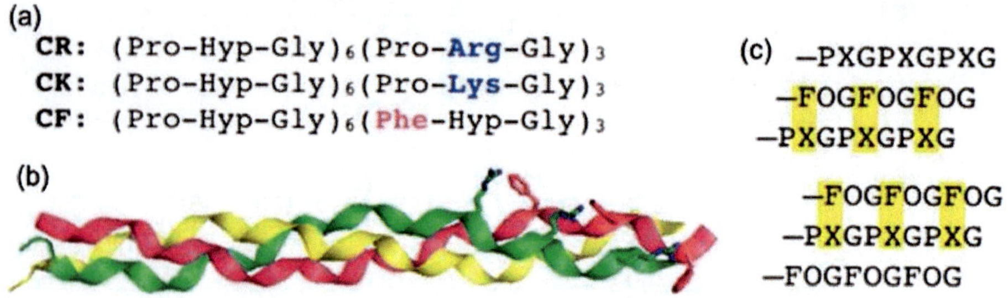

Fig. 3.19 **a** Sequences of the designed CMPs: CR, CK, and CF. **b** Illustration of the potential C-terminal cation-π interacting pairs in two AAB-type heterotrimeric arrangements, where X is Arg or Lys and O is Hyp. **c** A simple structure model showing the relative position of Arg (in green) and Phe (in magenta) in the C-terminal region of two different chains. The model was generated with PyMOL v0.99 and the PDB entry 1CAG. Reprinted with permission from [14]. Copyright 2017 American Chemical Society

initiation of UV-protective signaling (Fig. 3.20a). The mechanism by which UVR8 senses UVB is still unclear. Wu [15]. and Christie [16] independently reported the crystal structure of UVR8 and revealed the symmetric homodimeric structure of the 7-blade β-propeller (Fig. 3.20b). Each donut-shaped UVR8 monomer has a diameter of 40−50 Å and a height of 35 Å, with a water-filled tunnel in the center.

Detailed analysis of the UVR8 structure revealed the presence of an extensive network of intramolecular cation-π interactions in blade 6 (Fig. 3.20c). At the center of the network, Arg286 is surrounded by four aromatic amino acid residues, and the guanidinium group of Arg286 is located approximately 3.8 Å and 4.2 Å away from the indole rings of Trp302 and Trp285, respectively, allowing for strong cation-π interactions. Arg286 also interacts with the benzene ring of Tyr253 and binds weakly

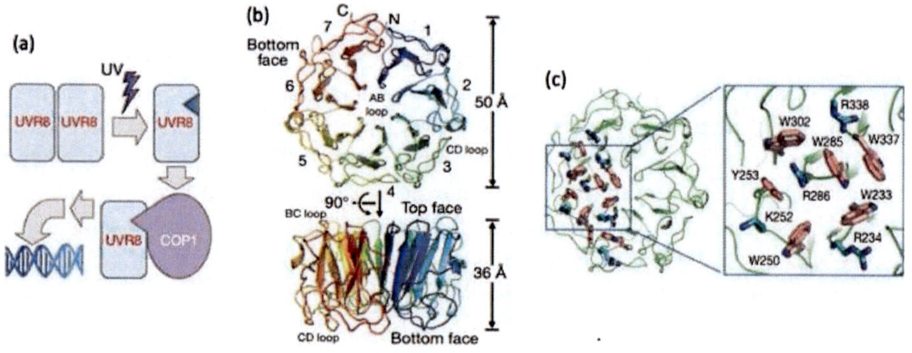

Fig. 3.20 **a** A schematic diagram of ultraviolet-B-induced, UVR8-mediated signaling cascade. Upon receiving ultraviolet-B (UVB) radiation, the UVR8 homodimer dissociates into monomers. The UVR8 monomer then associates with COP1, ultimately resulting in the activation of ultraviolet-B-responsive genes. **b** Overall structure of the UVR8 core domain. **c** A close-up view on the putative ultraviolet-B-sensing residues. Arginine and tryptophan residues are differentially colored. Reprinted by permission from Springer Nature [15], Copyright Springer Nature Singapore Pte Ltd. 2012

to Trp250. In addition, the indole ring of Trp285 binds to the guanidinium group of Arg338 via a cation-π interaction, while the indole ring of Trp285 binds to Trp233 and Trp337 via a π stacking interaction, forming an aromatic triangle. Furthermore, Trp233 is close to Arg234 through a cation-π interaction (Fig. 3.20b, c). It is rare to have such a high density of cation-π interactions in a protein (Fig. 3.20c).

To clarify the role of Trp285, UVR8 variants W285F and W285A were prepared and compared their optical properties. While wild-type UVR8 emitted at 335 nm at a UVB excitation wavelength of 295 nm, the variants W285F and W285A completely lost the ability to perceive UVB. When the protein crystals are irradiated with UV light, cracks appear in the wild-type UVR8 (Fig. 3.21), indicating that the dimer was dissociated to monomers. In contrast, no clack appears in the crystals of W285F and W285A even after prolonged exposure to UVB, indicating a loss of UVB responsiveness. These observations suggest that Trp285 is an essential component of the chromophore.

A possible mechanism for dissociation of dimer by irradiation was proposed as shown in Fig. 3.22. The homodimer is stabilized by Arg286 and Arg338 by Asp and Glu residues of the neighboring UVR8 molecule through H-bond. Excitation of W285 by UVB irradiation destabilizes the cation-π interactions with Arg286 and Arg338, leading to conformational change, which break H-bond with Asp and Glu residues of neighboring UVR8 molecule to trigger dissociation of dimer. Another mechanism

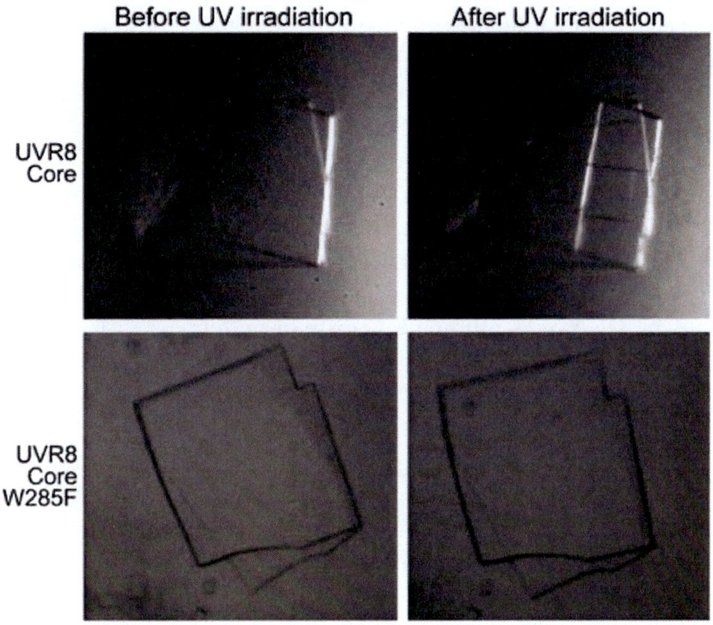

Fig. 3.21 UV radiation induced cracking of the crystals of the UVR8 core domain, but not the W285F crystals. Reprinted by permission from Springer Nature [15], Copyright Springer Nature Singapore Pte Ltd. 2012

for UVB-induced dissociation of the UVR8 homodimer has been proposed by Christie et al. [16].

3.3 Protein–DNA Association

3.3.1 PDB Analysis of Protein–DNA Complexes

Gromiha et al. analyzed the effect of cation-π interaction on the stability of 62 protein–DNA complexes extracted from PDB by using the Amber force field [17]. By this analysis, the presence of cation-π interactions was observed in 92% of the complexes. Table 3.7 shows the number and percentage of cationic amino acid residues Lys and Arg and four different bases in 62 protein–DNA complexes and the number and percentage involved in cation-π interactions. Comparing Lys and Arg residues, the percentage of Arg interaction was about twice as high as that of Lys, although there was no difference in the percentage of Arg present (about 8%). As for the bases, the ratio of A/T to G/C is found to be comparable. The aromatic chains with purine backbone (A and G) show a high frequency of cation-π interaction, and guanine is involved at a higher frequency than adenine. On the other hand, pyrimidine (C and T) is hardly involved in the interaction.

Table 3.8 shows that the average interaction energies for Lys/A and Arg/A are −2.01 and −2.10 kcal/mol, respectively, and that of Lys/G is −2.27 kcal/mol, which is stronger than that of Lys/A. Surprisingly, the Arg/G pair has the strongest interaction energy of −4.31 kcal/mol, the interaction with C is smaller than that with A and G, and the interaction with T is positive.

A typical stable structure of the Arg/G pair from calculations is shown in the Fig. 3.23, where the CZ of Arg is close to the five-membered ring part of G. This indicates that the arrangement occurs so that the five-membered ring with high electron density effectively interacts with Arg residue. These results will help us to understand the DNA recognition mechanism of proteins.

3.3.2 Nuclear Cap-Binding Complexes

The heterodimeric nuclear cap-binding complex (CBC) specifically recognizes the monomethylguanosine 5’ cap structure of eukaryotic RNA polymerase II transcripts, such as pre-mRNA and m^{7}GpppG (Fig. 3.24). CBC is a heterodimer composed of two polypeptides, CBP80 and CBP20, which are highly conserved in all eukaryotes. This RNA/CBC complex plays an important role in pre-mRNA splicing and mRNA nuclear export.

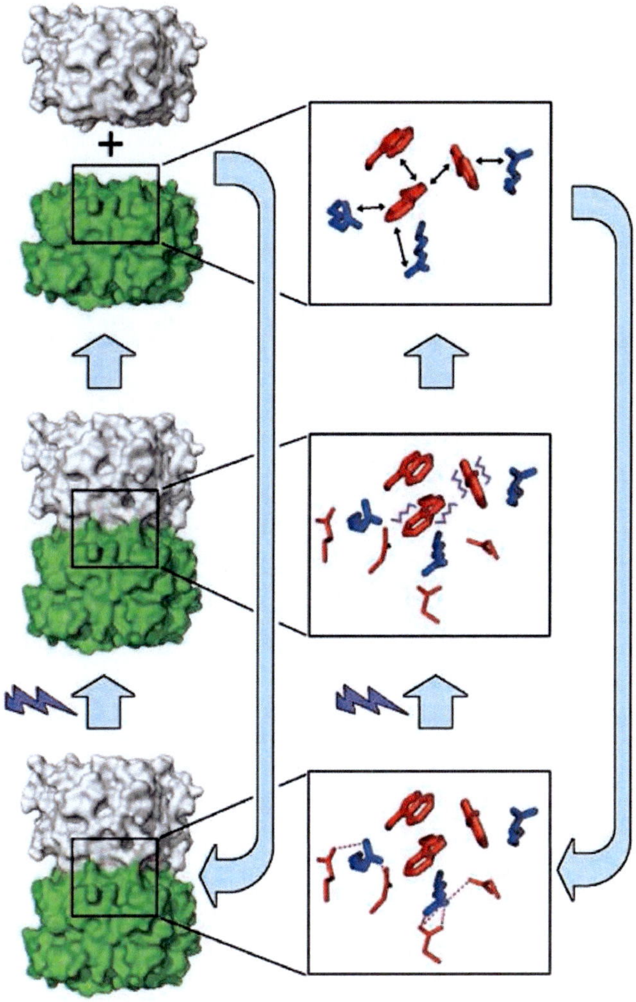

Fig. 3.22 A proposed mechanism for ultraviolet-B sensing by UVR8. Reprinted by permission from Springer Nature [15], Copyright Springer Nature Singapore Pte Ltd. 2012

Table 3.7 Composition and cation-π interactions exhibited by each amino acid/base in protein–DNA complexes

Amino acid/base	N[a]	Comp (%)[b]	$N_{cat\text{-}\pi}$[c]	$\%_{cat\text{-}\pi}$[d]
Lys	1014	7.92	148	14.60
Arg	1024	8.00	247	24.12
A	678	27.97	140	20.65
G	523	21.58	239	45.70
C	532	21.95	16	3.01
T	688	28.38	0	0.00

[a,b] N and Comp are, respectively, the number and composition of amino acid/base. [c,d] $N_{cat\text{-}\pi}$ and $\%_{cat\text{-}\pi}$ are, respectively, number and % of cation-π interactions in 62 protein–DNA complexes

Table 3.8 Average cation-π interaction energy combinations of amino acids and bases

Base	Lys (kcal/mol)	Arg (kcal/mol)
A	−2.01	−2.10
G	−2.27	−4.31
C	−1.52	−1.36
T	1.17	1.55

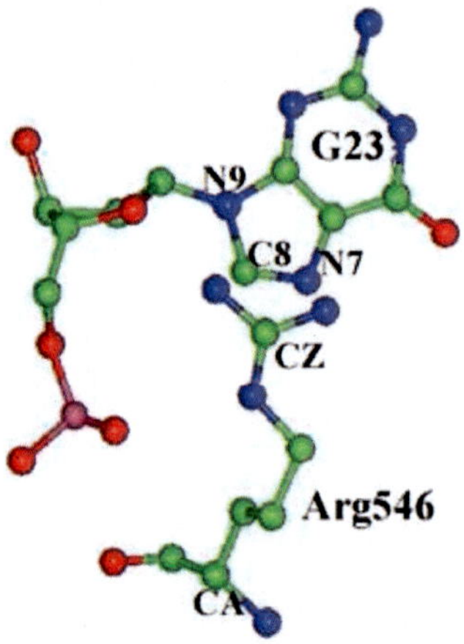

Fig. 3.23 Cation-π interaction between CZ of Arg546 and 5-member ring of G23 in the protein–DNA complex 1ign (PDB code). Reprinted from [17], Copyright 2004, with permission from Elsevier

In the X-ray structure of the m^{7}GpppG/CBC complex, a stacking structure was found in which two tyrosines Y43 and Y20 sandwich a positively charged *N*-methylguanosine (Fig. 3.25). This geometry strongly suggests significant contribution of synergistic π-cation-π interactions in this complex (see, Sect. 2.5.1). Furthermore, a stacking structure between the terminal guanosine and tyrosine Y138 has also

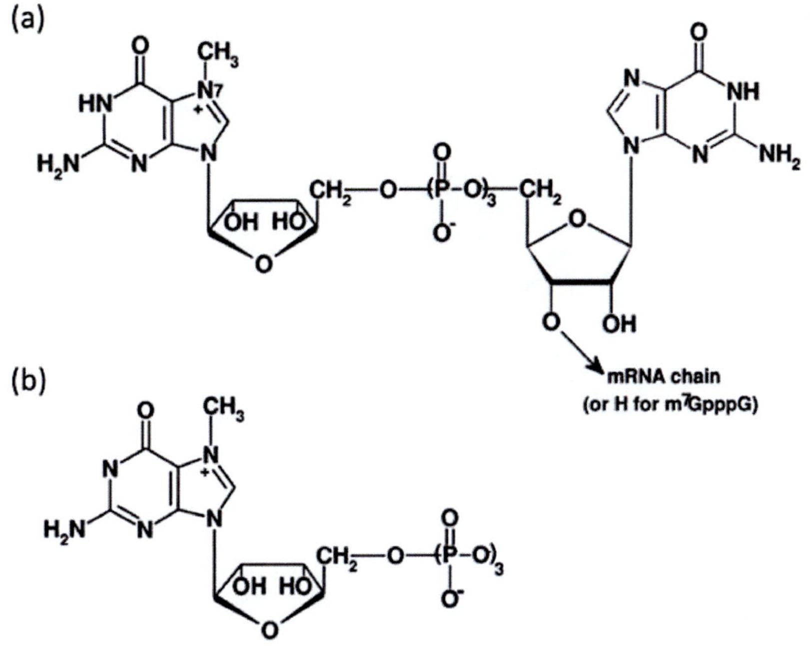

Fig. 3.24 Chemical structure of the monomethylated cap and its synthetic analogues used in the studies: **a** m^7GpppG (P1-7-methylguanosine-5′ P3-guanosine-5’-triphosphate), **b** m^7GTP (7-methylguanosine triphosphate). Reprinted from [18], Copyright 2009, with permission from Elsevier

been found. Such a sandwich structure is also found in other cap-binding proteins, e.g., CBP20 and eIF4E (eukaryotic initiation factor 4E) complexes.

To clarify that this characteristic sandwich structure is essential for CBC/cap recognition, the equilibrium association constants (K_{as}) of the CBC mutant, in which tyrosine is replaced by phenylalanine or alanine, and the cap analogues, m^7GpppG and m^7GTP (Fig. 3.24), were determined by fluorescence titration (Table 3.9) [18].

Table 3.9 summarized the equilibrium binding constants, K_{as}, and the Gibbs free energies of the corresponding bonds, $\Delta G°$, obtained from the titration experiments. Replacing the tyrosine with phenylalanine or alanine resulted in a significant decrease in the binding free energy, $\Delta G°$, of the high-affinity ligand m^7GpppG compared to the wild type. On the other hand, in the case of m^7GTP, only 0.9 and 0.25 kcal/mol reduction for Y20A CBC and Y20F CBC was observed, respectively. Thus, the affinity of m^7GTP for the Y20 CBC mutant was higher than that of m^7GpppG.

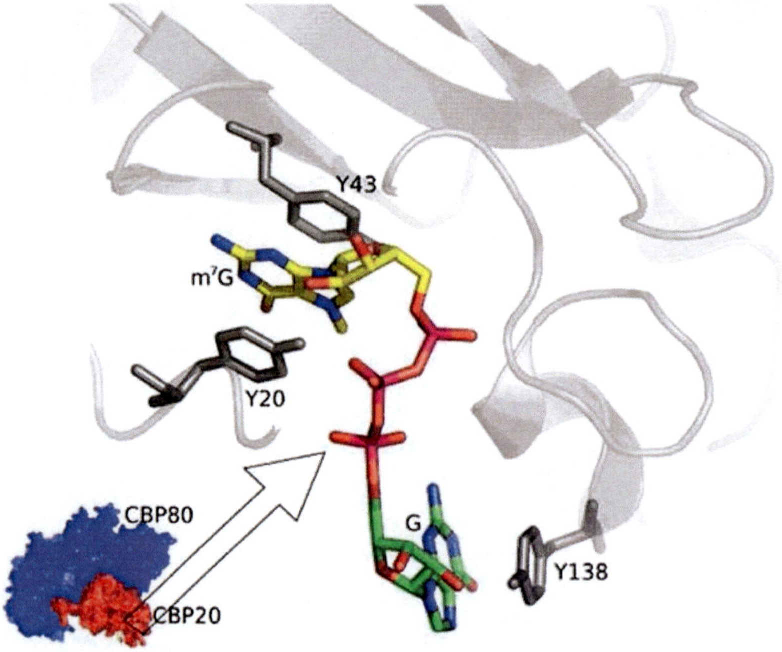

Fig. 3.25 View of the stacking interactions between m^7GpppG and CBC (PDB ID 1h2t) using PyMOL. Reprinted from [18], Copyright 2009, with permission from Elsevier

Table 3.9 Equilibrium association constants (K_{as}) and binding Gibbs free energies ($\Delta G°$) for the cap–CBC complexes at 20 °C

CBC mutant	m^7GpppG		m^7GTP	
	Kas($\times$106 M^{-1})	$\Delta G°$(kcal mol^{-1})	Kas($\times$ 106 M^{-1})	$\Delta G°$(kcal mol^{-1})
Wild type	249 $\pm$ 62	-11.3 ± 0.1	31.1 ± 6.6	-10.0 ± 0.1
Y138A	67 ± 15	-10.5 ± 0.1	18.8 ± 0.9	-9.75 ± 0.03
Y43F	66 ± 17	-10.5 ± 0.2	6.4 ± 1.0	-9.12 ± 0.09
Y43A	24 ± 2	-9.89 ± 0.04	7.0 ± 1.4	-9.2 ± 0.1
Y20F	1.7 ± 0.3	-8.35 ± 0.09	6.3 ± 0.9	-9.11 ± 0.08
Y20A	1.9 ± 1.8	-8.4 ± 0.5	18.8 ± 1.1	-9.75 ± 0.03

Considering the results of other kinetic experiments, a two-step binding mechanism has been proposed as follows. Y20 and Y43 are involved in the recognition of m^7G and are stabilized by cation-π interactions as shown in Fig. 3.26, and then, Y138 binds by π-π interaction with the next nucleotide.

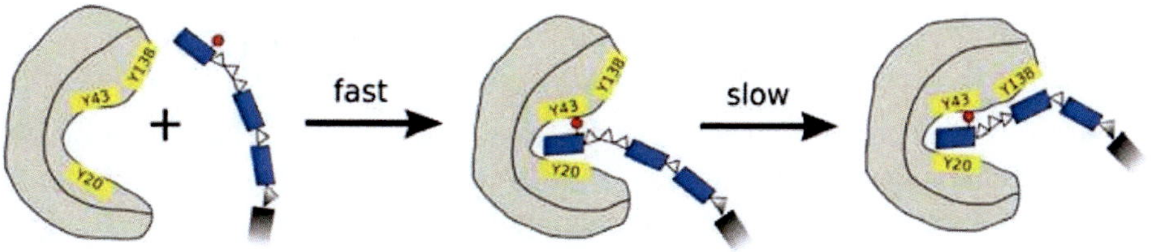

Fig. 3.26 Kinetic model of the CBC–mRNA association showing the first, faster, and the second, slower, step. Reprinted from [18], Copyright 2009, with permission from Elsevier

3.4 Protein–Ligand Association

Cation-π interactions have been known to play an important role in the recognition of ligands by proteins in enzyme-inhibitor, receptor-ligand, transporter-ligand, and antibody-hapten systems. There are two types of interactions in the recognition of ligands by proteins: (1) combinations of aromatic amino acid residues (Phe, Tyr, or Trp) on target proteins and positively charged ligands and (2) combinations of cationic amino acid residues and small molecule ligands with π-systems [19]. An example of the former is the case of quaternary ammonium-containing acetylcholine (ACh) specifically interacting with Tyr and Trp residues of the acetylcholine receptor (AChRs) (Fig. 3.27a). An example of the latter is interaction of the cationic amino acid residues of monoclonal antibody 4D5 with benzo[a]pyrene having a π-system (Fig. 3.27b). The former may be distinguished as cation-π interactions and the latter as π-cation interactions. This document does not use such a classification because there are a variety of cases where such a classification is not applicable.

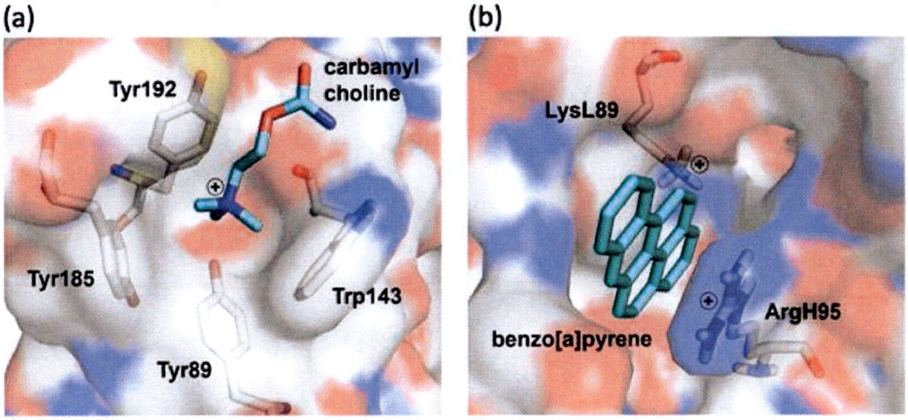

Fig. 3.27 Comparison of the cation-π and π-cation interactions in ligand–protein recognition. **a** Carbamylcholine binding to acetylcholine-binding protein. Model from PDB code 1UV6. **b** Benzo[a]pyrene binding to monoclonal antibody 4D5. Reprinted with permission from [19]. Copyright 2018 American Chemical Society

3.4.1 Recognition of Quaternary Ammonium Ions

Schärer et al. focused on the role of a cation-π interaction in the action of serine proteases in blood coagulation [20]. They studied the activity of tricyclic inhibitors **8** and **9** possessing a quaternary ammonium ion and an uncharged *tert*-butyl group, respectively, against thrombin, trypsin, and the serine proteases Factor VIIa and Factor Xa. The values of inhibition constant *K*i for them are shown in Fig. 3.28.

The onium salt **8** showed much higher activity against Factor Xa than the trimethyl form **9**. The X-ray crystallographic analysis of the cocrystals of Factor Xa and **8** (Fig. 3.29) suggested the preferential recognition of the onium ion by the receptor through cation-π interactions. The tricyclic framework is located in the center of the cavity, and the phenylamidinium residue enters the S1 pocket and interacts with the carboxylate side chain of Asp189. The quaternary ammonium ion of **8** binds to the aromatic box of the S4 pocket as predicted, and the distance between the closest carbon atom and the benzene ring is 4.6–5.0 Å, indicating the contribution of cation-π interactions.

Salonen et al. synthesized new tricyclic compounds **10** and **11** and found that compound **10** with chlorothiophenyl group showed high activity against Factor Xa (K_i = 9 nM) compared to that of *tert*-Bu derivative **11** (K_i = 550) [21]. The X-ray crystal structure for the complex of Factor Xa and **10** is shown in Fig. 3.30. The tricyclic skeleton is located in the center of the cavity, and the chlorothiophenyl group is in the S1 pocket, approaching Tyr228 by dipolar interaction. The quaternary ammonium ion is located in the center of the aromatic box at the S4 pocket composed by Trp215, Phe174, and Tyr99, indicating that it is effectively recognized by the cation-π interactions (Fig. 3.30). The distances between the nitrogen atom and the closest carbon atoms of Trp215, Phe174, and Tyr99 are 4.3 Å, 4.7 Å, and 4.8 Å, respectively. The difference in binding affinities between **10** and **11** was ΔG = 2.5 kcal/mol, corresponding to 0.8 kcal/mol per benzene ring.

There are a number of similar examples involving the complexes of quaternary ammonium ions with aromatic boxes in a variety of protein–ligand complexes, some of which are shown in Fig. 3.31a–d. In addition, examples of the complexes between the aromatic boxes and ammonium ions generated by protonation of amino groups are

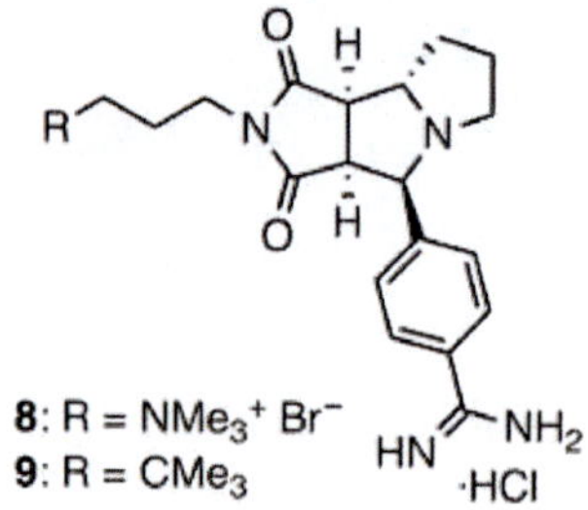

Inhibitor	K_i [μM] thrombin	trypsin	Factor VIIa	Factor Xa
8	7.76	10.5	> 69	0.28
9	0.13	7.1	> 69	29

Fig. 3.28 Activities (inhibition constants *Ki* [μM]) of inhibitors (±)-**8** and (±)-**9**

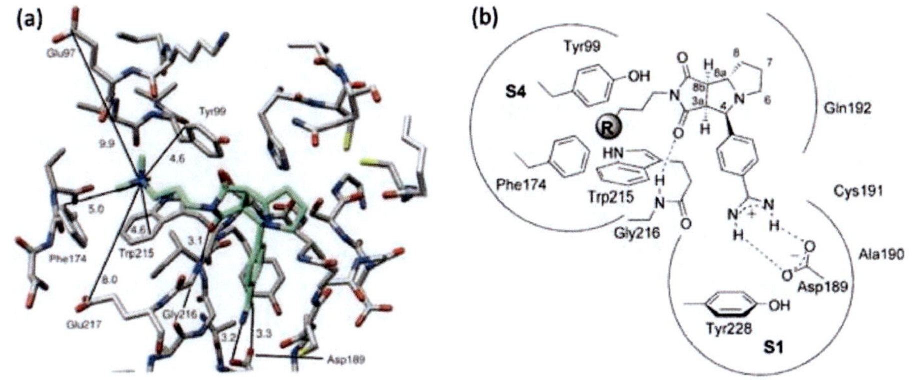

Fig. 3.29 **a** Binding mode of inhibitor (3a*S*,4*R*,8a*S*,8b*R*)-**8** in the active site of Factor Xa according to X-ray crystal-structure analysis. Interatomic distances are in Å (Reproduced from Schärer et al. 2005). **b** Schematic representation of the active site of Factor Xa complexed with (3a*S*,4*R*,8a*S*,8b*R*)-**8**. From [21] Copyright © 2009 by John Wiley Sons, Inc. Reprinted by permission of John Wiley & Sons, Inc.

Fig. 3.30 **a** Molecular structures of **10** and **11** with K_i values against Factor Xa and thrombin. **b** X-ray crystal structure of **10** in the complex with Factor Xa (PDB code: 2JKH). From [21] Copyright © 2009 by John Wiley Sons, Inc. Reprinted by permission of John Wiley & Sons, Inc.

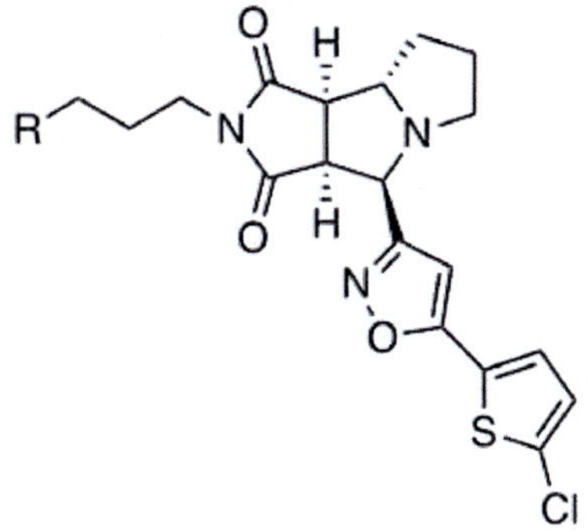

10: R = NMe_3^+ Br^- — *Ki*(FXa) = 550 nM — *Ki*(Thr) = 17800 nM

11: R = CMe_3 — *Ki*(FXa) = 9 nM — *Ki*(Thr) = 35100 nM

shown in Fig. 3.32. These examples clearly show that the recognition of ammonium ions by the aromatic boxes is universal in protein–ligand interactions.

3.4.2 Biosynthesis of Aristolochene

The aristolochene synthase (PR-AS) from Penicillium roqueforti catalyzes the formation of bicyclic sesquiterpenes aristolochene (**15**) and epi-aristolochene by magnesium-dependent cyclization of farnesyl diphosphate (**12**) and subsequent

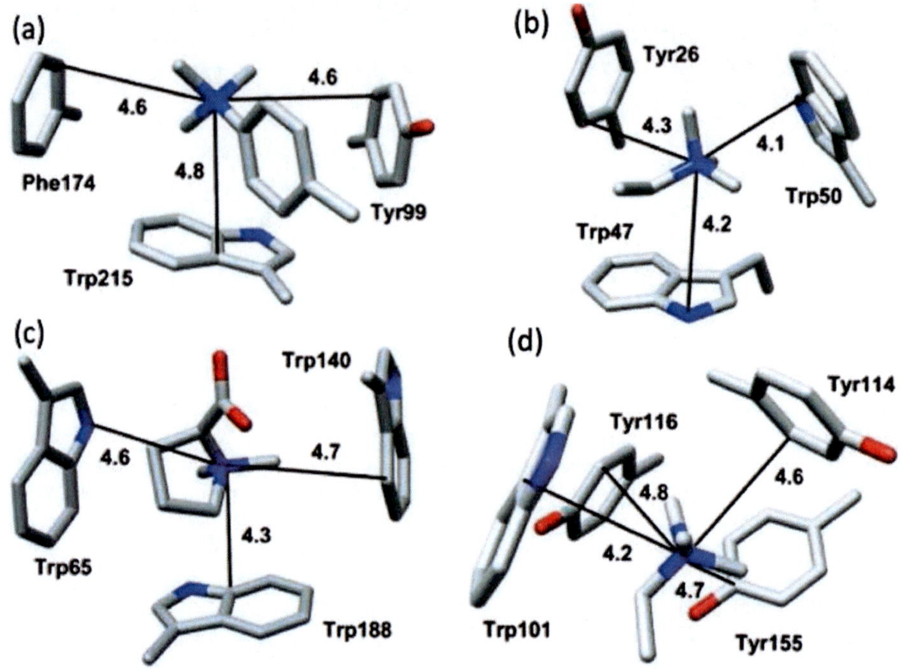

Fig. 3.31 Examples from PDB searches showing quaternary ammonium ions bound in the S4 pocket of Factor Xa **a** and in similar aromatic boxes of other proteins **b-d**. The shortest ligand–protein distances are indicated in Å; red O, blue N, and gray C. PDB-codes: **a** 1LPK, **b** 1PFB, **c** 1R9Q, and **d** 1LN3. From [20] Copyright © 2005 by John Wiley Sons, Inc. Reprinted by permission of John Wiley & Sons, Inc.

cationic rearrangement (Fig. 3.33). However, it is not clear how the high-energy cationic intermediates are stabilized in all terpene synthase-dependent cyclizations.

On the basis of the X-ray crystal structure of aristolochene synthetase, Allemann et al. hypothesized that when farnesyl diphosphonate (FDP) is incorporated into the active site, Trp334 stabilizes the intermediate cation through a cation-π interaction (Fig. 3.34) [22]. To confirm this hypothesis, the enzyme activity and the extent of accumulation of the intermediate germacrene A (**13**) were investigated using Trp334 mutant enzyme, as they can be used as indicators of the stabilization of eudesmane cation (**14**). The catalytic activity was investigated using mutants W334F, W334L, and W334V, in which tryptophan residue Trp334 was replaced by phenylalanine, leucine, and valine, respectively. Table 3.10 summarizes the results of the effect of mutation on the enzyme activities, in which the activity of the phenylalanine mutant is similar to that of wild type, but leucine and valine mutants showed a large decrease in k_{cat} and k_{cat}/K_M values, accumulation of germacrene A (**13**), and a significant decrease in aristrochene (**15**) (Table 3.10). These results indicate that the aromatic ring of Trp334 contributes to the stabilization of the intermediate eudesmane cation (**14**).

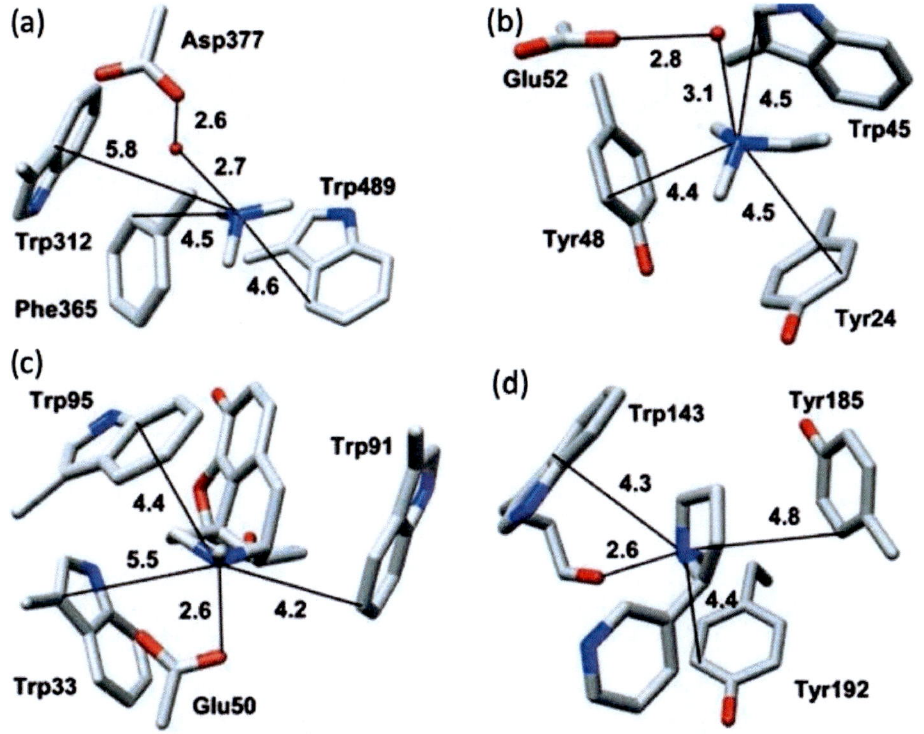

Fig. 3.32 Examples from PDB searches showing protonated tertiary ammonium ions bound in aromatic boxes. The shortest ligand–protein distances are indicated in Å; red O, blue N, and gray C. PDB-codes: **a** 1H37, **b** 1KNA, **c** 1Q0Y, and **d** 1UW6. From [20] Copyright © 2005 by John Wiley Sons, Inc. Reprinted by permission of John Wiley & Sons, Inc.

Moreover, the effect of the electron density of the aromatic ring on the amount of aristolochene produced was investigated by replacing Trp 334 with para-substituted phenylalanine. As shown in the Fig. 3.35, there was a good correlation between the interaction energy and the amount of aristolochene. This suggests that Trp 334 contributes to the stabilization of the transition state of cyclization and reaction intermediates through cation-π interactions for the conversion of **13** to **15**.

In addition, Faraldos and Allemann also found that iminium ion **16**, which mimics the eudesmane cation (**14**), acts as an inhibitor of aristolochene synthase [23]. This fact also supports the role of Trp334.

3.4.3 Cellulase–Lignin Association

The enzymatic hydrolysis of cellulose in lignocellulose to glucose is one of the most important steps in the production of biofuels. However, as the adsorption of lignin on expensive cellulases is non-productive and greatly hampers the development of

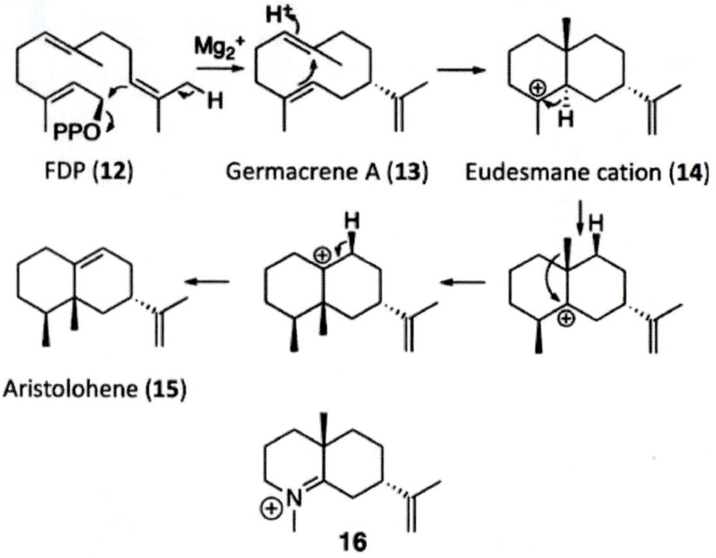

Fig. 3.33 Biosynthesis of (+)-aristolochene **15** and inhibitor **16**. Reprinted with permission from [22]. Copyright 2011 American Chemical Society

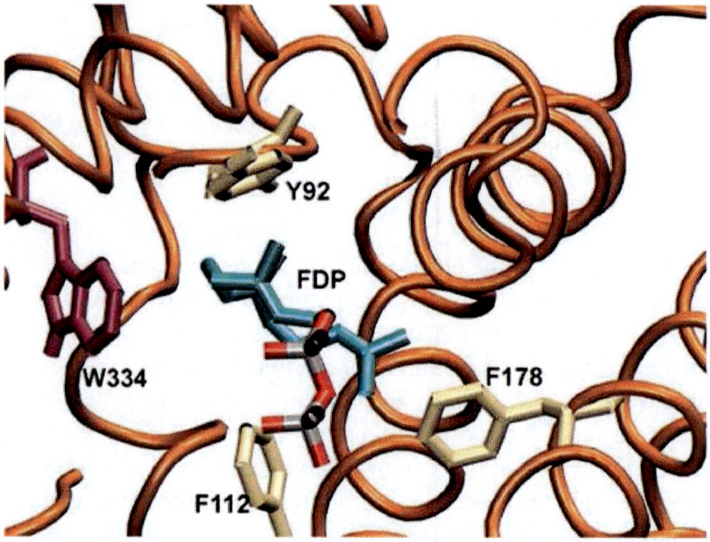

Fig. 3.34 Sketch of the active site of PR-AS, indicating the position of Trp334 and a modeled conformation of FDP (PDB code: 1F1P). Reprinted with permission from [22]. Copyright 2011 American Chemical Society

Table 3.10 Kinetic parameters and product distributions for WT PR-AS and PR-AS-W334 natural mutants

Enzyme	k_{cat} (min^{-1})	K_M (μm)	k_{cat}/K_M (min^{-1} μM^{-1})	**15** (%)	**13** (%)
WT	5.04	0.53	9.51	92	8
W334F	0.38	0.28	1.40	82	18
W334Y	0.43	1.03	0.42	62	38
W334H	0.02	0.13	0.14	10	90
W334L	0.16	0.75	0.22	2	98

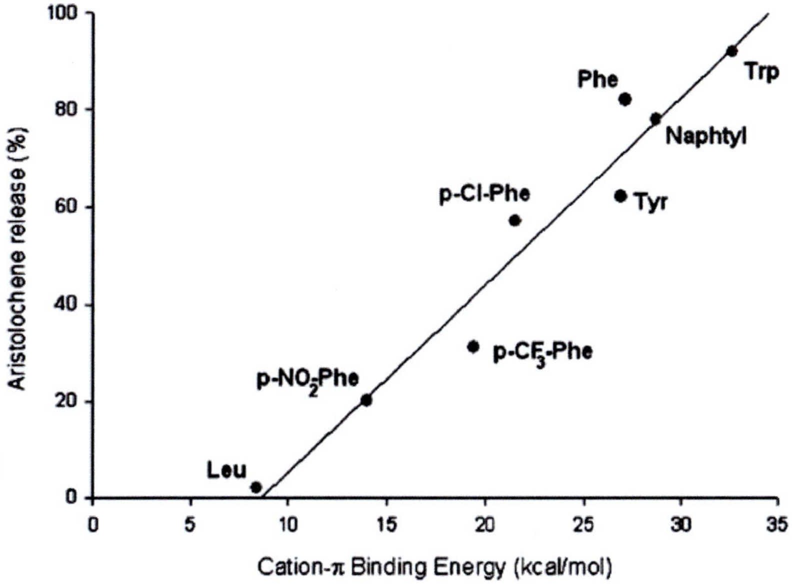

Fig. 3.35 Relationship between the cation-π binding energies and the percentage of aristolochene produced (r = 0.95). Reprinted with permission from [22]. Copyright 2011 American Chemical Society

biorefinery, elucidation of the cellulase–lignin interaction mechanism is important in reducing such a non-productive effect.

Zheng et al. directly measured the interaction forces between cellulase and substrates (cellulose and lignin) in aqueous solution and at the nanoscale using surface force apparatus (SFA) [24]. The results showed that the adhesion energy of lignin–cellulase was 2.85±0.12 mJ/m^2, whereas that of cellulose–cellulase was 1.42±0.67 mJ/m^2 (Fig. 3.36a). The strong adhesion of cellulase to lignin confirms that cellulase molecules are more easily adsorbed on lignin than on cellulose.

With increases in the Na^+ concentration (1−100 mM), the adhesion energy slightly decreased from ~2.47 to ~2.25 mJ/m^2. On the other hand, K^+ had a stronger blocking effect than did Na^+. As the concentration of K^+ in the CH_3CO_2H/CH_3CO_2Na (AA/SA) buffer was increased from 1 to 100 mM, the adhesion energy between the

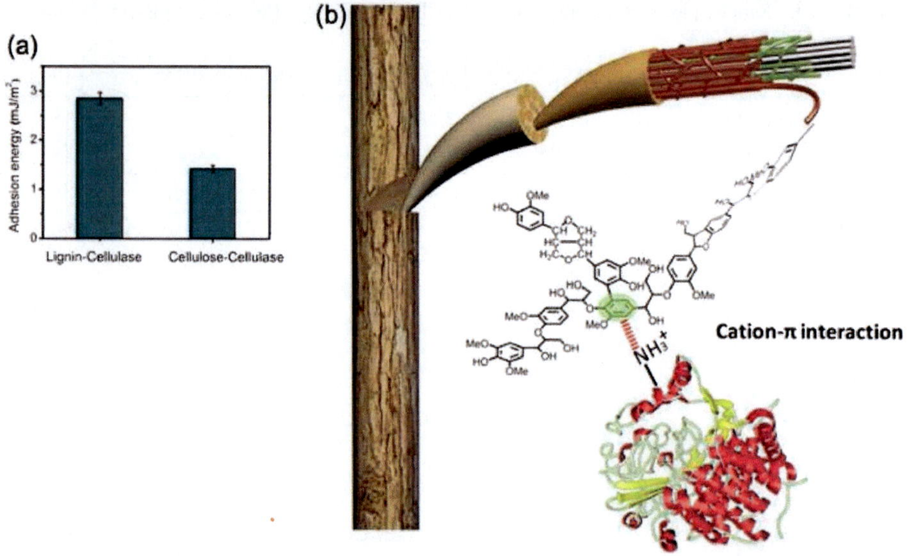

Fig. 3.36 **a** Maximum normalized adhesion energies of lignin−cellulase and cellulose−cellulase. **b** Schematic diagram of the cation-π interaction between lignin and cellulase. Reprinted with permission from [24]. Copyright 2021 American Chemical Society

lignin and cellulase layers decreased markedly from ~2.54 mJ/m^2 to ~1.02 mJ/m^2. The dramatic decrease in adhesive strength with the addition of K^+ is mainly attributed to the fact that K^+ forms a strong cation-π pair with the lignin surface, occupying the cationic sites on the lignin and reducing the cation-π interaction between cellulase and lignin as shown in Fig. 3.36b. These findings may provide nanomechanical insights into the enhancement of enzymatic hydrolysis efficiency.

3.4.4 *Recognition of Purine Analogues*

Adenophostin A (AdA)(**18**) is a potent agonist of the D-*myo*-inositol 1,4,5-trisphosphate receptor (Ins(1,4,5)P_3R) and is known to promote Ca^{2+} release. Potter et al. synthesized 2-aminopurine analogues of AdA (**19–23**) (Fig. 3.37) and investigated their activities and the structures of the complexes with (Ins(1,4,5)P_3R) [25]. Table 3.11 showed that each of these analogues (**19–23**) was a complete agonist of equal or slightly lower potency than AdA. On the other hand, from another experiment, the activity of 2-methoxypurine analogue **19** is known to be 5.9 times less potent than ADA, suggesting the importance of the 2-amino group for the activity. The nearly equal activities of **20** and **21** suggest that the contribution of 6-NH_2 of AdA to Ins(1,4,5)P_3R recognition is negligible.

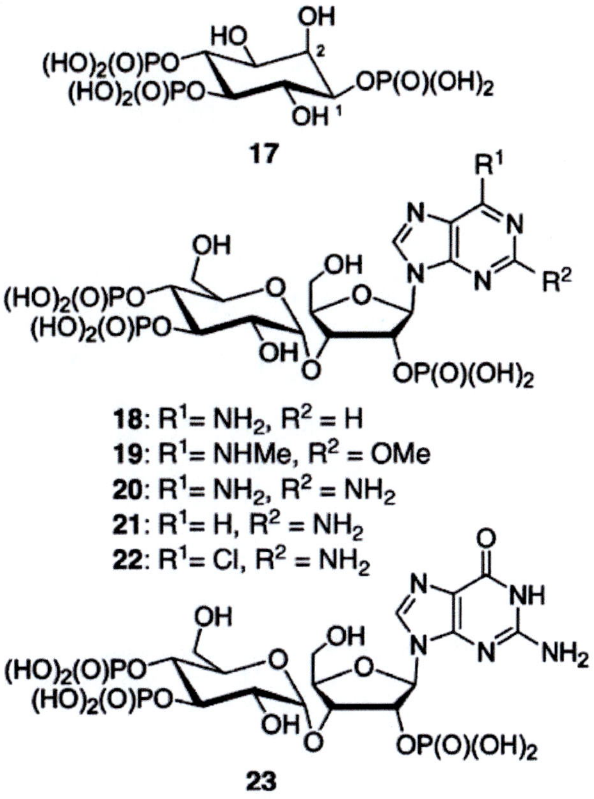

Fig. 3.37 Chemical structures of D-*myo*-inositol 1,4,5-trisphosphate, ADA, and ADA analogues

Table 3.11 Stimulation of Ca^{2+} Release via Recombinant Rat Type 1 Ins(1,4,5)P_3 Receptor

Substrate	EC_{50} (nM)[a]	Hill coefficient	% release[b]
Ins(1,4,5)P_3 (**17**)	23.7±3.7	1.30 ±0.14	81±1
adenophostin A (**18**)	0.9±0.1	1.20±0.06	81±2
2,6-diaminopurinophostin (**20**)	2.2±0.5	1.32±0.10	80±2
2-aminopurinophostin (**21**)	2.4±0.8	1.00±0.06	80±4
chlorophostin (**22**)	3.2±1.0	1.01±0.02	85±1
guanophostin (**23**)	1.3±0.3	0.99±0.08	83±1

[a] The half-maximally effective concentration (EC_{50}). [b] The percentage of the Ca^{2+} stores released by a maximally effective concentration of each ligand are shown

The docking simulations between Ins(1,4,5)P_3R (PDB file number 1N4K) binding nuclei and AdA derivatives revealed that the highest-score binding mode of these new ligands closely resembled that of AdA. Figure 3.38 shows the structure of the complex of substrate **20** and Ins(1,4,5)P3R. As can be seen from the figure, the guanidinium side chain of Arg504 is very close to the purine ring, suggesting the presence of a cation-π interaction. The lower potency of chlorophostin **22** and higher potency

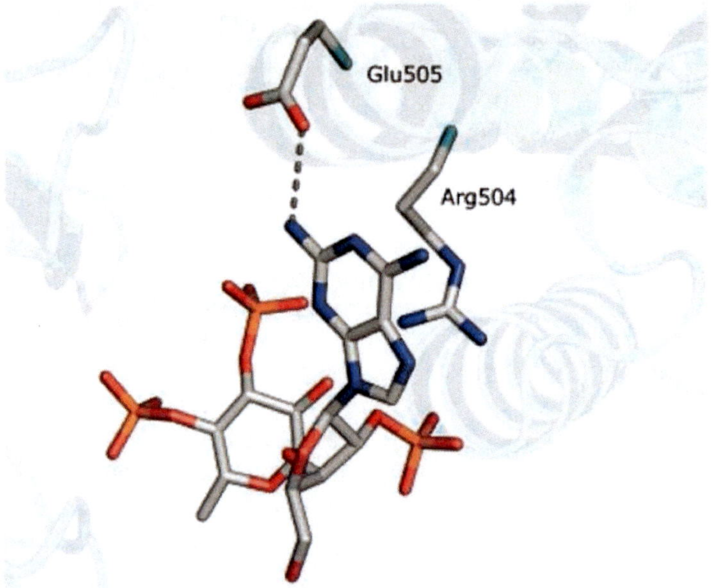

Fig. 3.38 Highest scoring docked conformation of 2,6-diaminopurinophostin **20** forms a hydrogen bond with the carboxylate of Glu505 (hydrogen atoms are not shown). Reprinted with permission from [25]. Copyright 2008 American Chemical Society

of guanophostin **23** in assays of Ca^{2+} release via recombinant Ins(1,4,5)P_3R are consistent with the model suggesting a cation-π interaction between the guanidinium side chain of Arg504 and the purine ring as shown in Fig. 3.38.

3.5 Ion Channels

Ion channels are small pores composed with proteins in the cell membrane that open and close in response to stimuli through which ions can pass. There are two types of ion channels: ligand-gated ion channels (LGICs), which open and close when a neurotransmitter binds to them, and voltage-gated ion channels (VGICs), which open and close by sensing the potential difference between the inside and outside of the membrane. When a ligand binds to a ligand-dependent ion channel, the channel in the receptor immediately opens. As a result, ions permeate according to the electrochemical gradient in and out of the cell and modulate the excitability of the postsynaptic membrane. There are many types of neurotransmitters known, such as acetylcholine, amino acids, and monoamines. Voltage-dependent channels exist in the cell membranes of nerves and muscles, are activated by depolarization of the cell membrane, and permeate Na^+, K^+, and Ca^{2+}. When these channels are blocked, the excitation of motor nerves and skeletal muscles is suppressed, resulting in paralysis.

3.5.1 Ligand-Gated Ion Channels

The pentameric receptors of the Cys-loop superfamily are known to be representative of LGICs. They are usually composed of five protein subunits: two alpha subunits and three beta, gamma, or delta subunits. The name "Cys-loop" comes from the fact that this type of receptor has a characteristic loop formed by 13 highly conserved amino acids between two cysteine (Cys) residues that form a disulfide bond near the N-terminal extracellular domain. These include excitatory cation-selective channels opened by acetylcholine (nAChR) and serotonin (5-HT_3) and inhibitory anion-selective channels opened by gamma-aminobutyric acid (GABA) and glycine.

Figure 3.39 shows the X-ray crystal structure of the human nicotinic acetylcholine α4β2 receptor (nAChR) [26]. The pentameric structure is formed by the combination of two subunits with α4 and β2. Each subunit consists of an N-terminal extracellular domain with four transmembrane α-helices and a large β-sheet. Ligand-binding sites are found at the subunit interface of the extracellular domain, in which nicotine is shown in red (Fig. 3.39). Dougherty et al. have shown that in the Cys-loop receptor, the cation-π interaction between ligand and receptor plays an important role in ligand recognition.

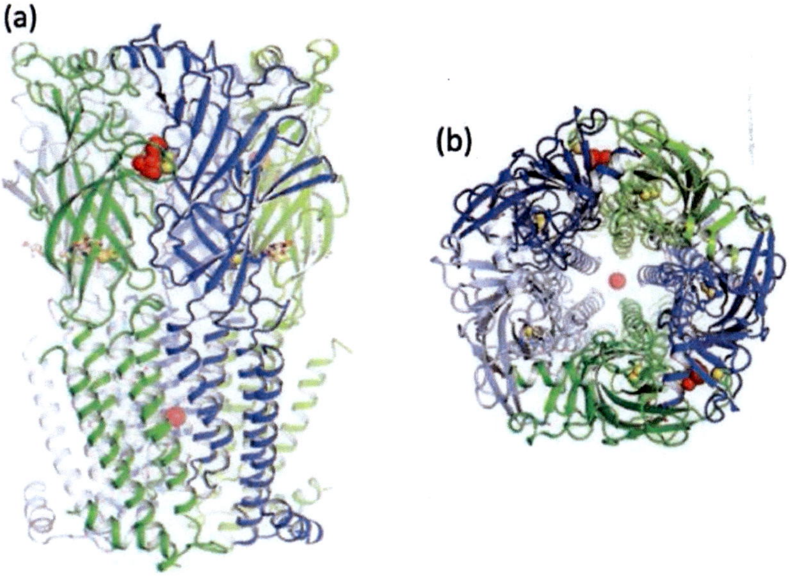

Fig. 3.39 Architecture of the α4β2 nicotinic receptor. **a** View parallel to the plasma membrane. α4 subunits are in green and β2 in blue. Nicotine (red) and sodium (pink) are represented as spheres. The Cys-loop and loop C disulfide bonds are shown as yellow spheres. N-linked glycans (brown) are shown as sticks. **b** View perpendicular to the plasma membrane looking from the extracellular side. Reprinted by permission from Springer Nature [26], Copyright Springer Nature Singapore Pte Ltd. 2016

3.5.1.1 Nicotinic Acetylcholine Receptors

Nicotinic acetylcholine receptors (nAChRs) are the best-studied Cys-loop receptors, and α4β2 receptors are one of the major nAChR subtypes expressed in the brain. It has been established that it binds with high affinity to nicotine and plays an important role in nicotine dependence from smoking. Each binding site consists of six "loops" labeled A-F. A-C form the "main face," and D-E provided by adjacent subunits form the "complementary face." At the binding site, aromatic residues are assembled to form an "aromatic box." These aromatic residues are named such as TyrA, TrpB, TyrC1, TyrC2, and TrpD, in such a way that they belong to which loop. Other Cys-loop receptor binding sites are also rich in Phe, Tyr, and Trp, and their assembled aromatic boxes are universally found.

As shown in Fig. 3.40, [27] the cationic moiety of nicotine is directed toward the main face of the binding site, and three non-covalent interactions are observed: (1) cation-π interaction with TrpB, (2) cation-π interaction with TyrC2, and (3) hydrogen bonding between the agonist NH^+ group and the CO group of TrpB backbone. In addition, hydrogen bonds via H_2O with the pyridine nitrogen of the agonist are found in the complementary subunits.

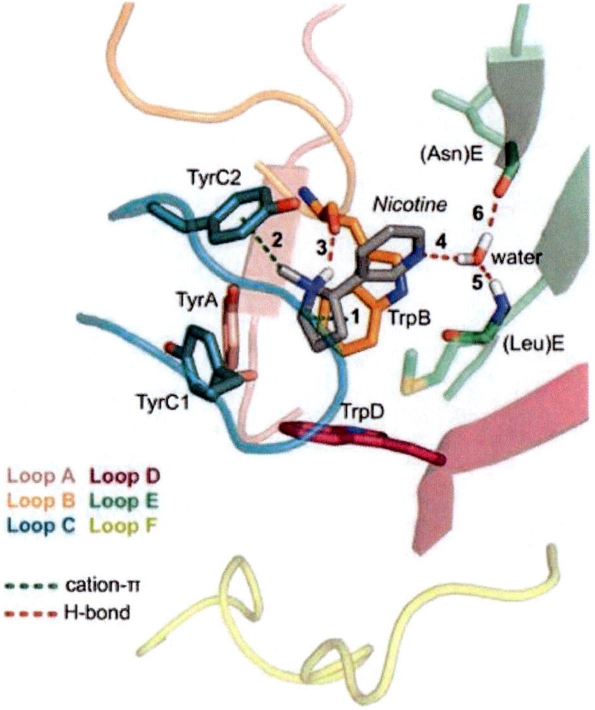

Fig. 3.40 Crystal structure of AChBP in complex with nicotine (PDB code 1UW6). Reprinted with permission from [27]. Copyright 2014 American Chemical Society

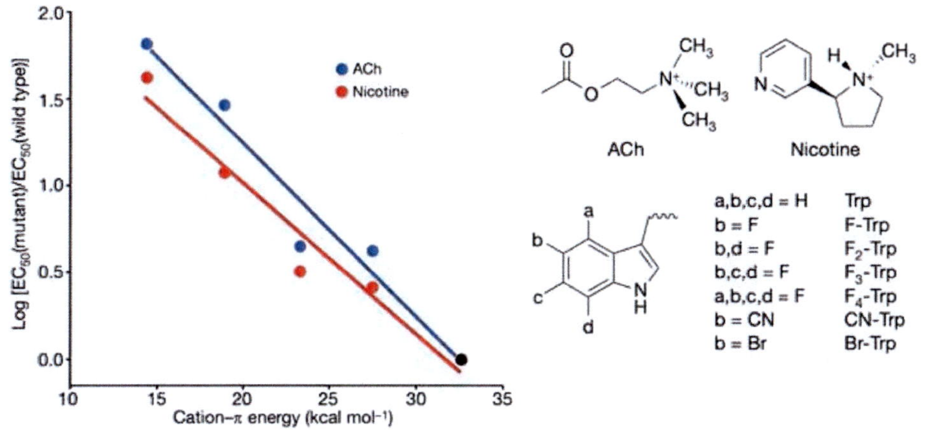

Fig. 3.41 Fluorination plots for α4β2 receptor with structures of ACh, nicotine, and unnatural amino acids considered here. An a, b, c, or d group is H if not indicated. Reprinted by permission from Springer Nature [28], Copyright Springer Nature Singapore Pte Ltd. 2009

The contribution of the cation-π interactions between ACh and nicotine with TrpB was demonstrated by the fluorination plots [28]. Figure 3.41 shows fluorination plots, in which the interaction energy decreased as the number of fluorine substitutions increased, and there was a good correlation between the EC_{50} and the interaction energy plots. This proves that the complexes with both ACh and Nicotine are stabilized by the cation-π interaction with TrpB. Moreover, in the case of cytidine ligands, TyrC2 also contributes to stabilization through cation-π interactions by similar plots with the fluorinated form of tyrosine [29].

3.5.1.2 5-HT$_3$ Receptors

Serotonin type 3 receptor (5-HT$_3$R) is a ligand-dependent ion channel present in the central and peripheral nervous system. 5-HT$_3$R is a therapeutic target, and ondansetron and granisetron are used in clinical practice to inhibit receptor activity. The antiemetic compounds for which they were formulated are commercially available as Zofran and Kytril, respectively. It is proposed that the cation-π interaction of W183 in the ligand-binding pocket with these inhibitors promotes their binding as shown in Fig. 3.42 [30].

3.5.1.3 GABA and Glycine Receptors

GABA$_A$ (γ-aminobutyric acid) and glycine receptors are anion-selective Cys-loop receptors that mediate fast inhibitory neurotransmission in neurons. These receptors have high homology and 30–35% sequence identity to the glutamate-gated chloride channel (GluCl) receptor of the nematode C. elegans, making this structure a useful

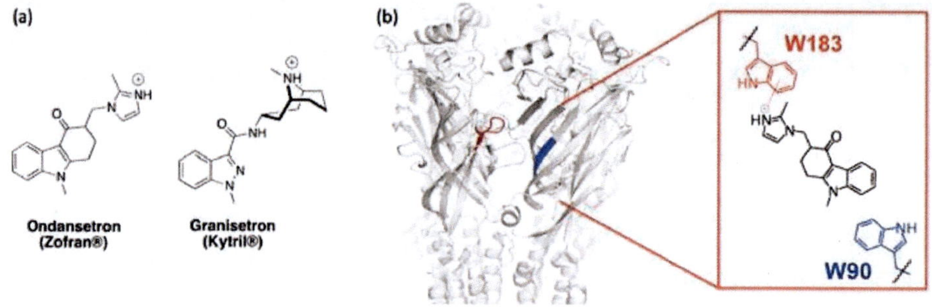

Fig. 3.42 **a** Chemical structures of drugs. **b** Detail of agonist binding site. Reprinted with permission from [30]. Copyright 2012 American Chemical Society

template for the binding sites of these receptors. In the GluCl receptor, aromatic residues Phe, Tyr, and Tyr are found in the A, B, and C2 sites of the aromatic box of nAChR, respectively. The structure shows strong cation-π interaction between glutamate and TyrB as well as TyrC2 (Fig. 3.43).

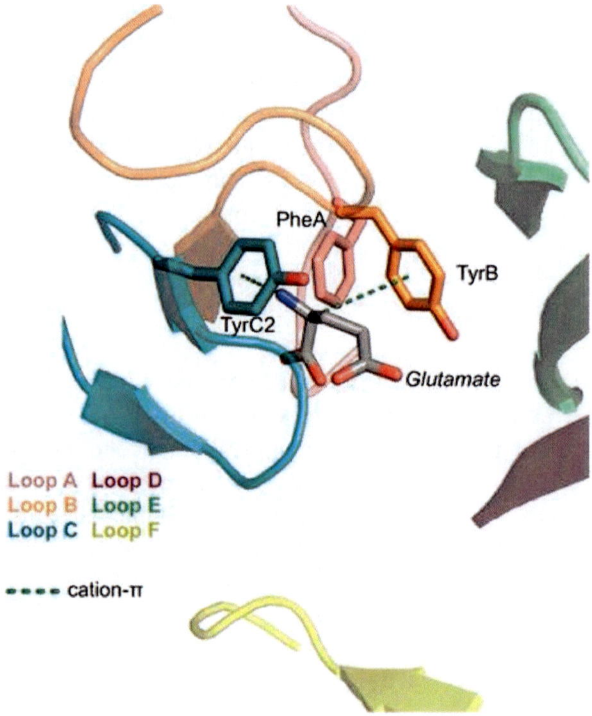

Fig. 3.43 Ligand-binding site of C. elegans GluCl in complex with glutamate (PDB code 3RIF). The cationic group of glutamate participates in a cation-π interaction with the conserved aromatic box site TyrC2 and the TyrB site as well. The carboxylates of glutamate are paired with Arg side chains from the complementary face of the binding site. Reprinted with permission from [27]. Copyright 2014 American Chemical Society

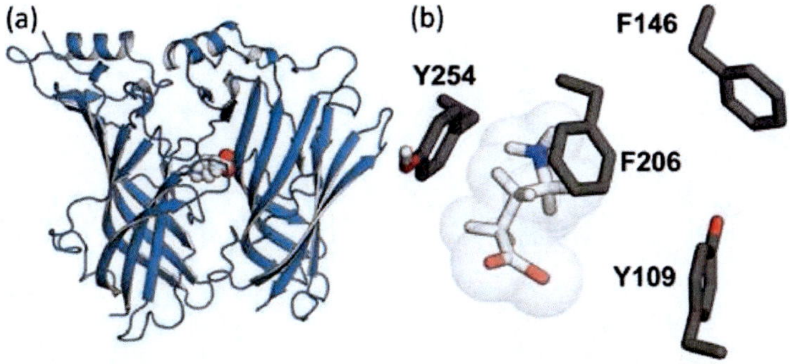

Fig. 3.44 **a** Homology model of the extracellular domain of the RDL receptor based upon the crystal structure of AChBP with carbamylcholine bound (PDB ID; 1uv6). Only two of the five subunits have been shown for ease of viewing. The ligand-binding site is located between the two subunits. **b** Enlargement showing GABA docked into the binding site and surrounded by the "aromatic box" residues

GABA-activated Cys-loop receptors shown in Fig. 3.44 are also present in invertebrates, and they are targets for a variety of pesticides. To clarify the role of the aromatic box, the receptor binding pocket incorporating a series of fluorinated Phe derivatives was used for the investigation of the interaction with GABA [31]. The cation-π-binding capacity of both Phe206 (loop B) and Tyr254 (loop C) to GABA showed a clear correlation with EC_{50}. These results show that the two aromatic residues (loops B and C) are positioned to contribute to the interaction with the primary ammonium cation of GABA through cation-π interactions, which provide insights for the design of insecticides that act on insect GABA receptors.

3.5.2 *Voltage-Gated Ion Channels*

Potential-dependent ion channels are transmembrane proteins that are activated by changes in the membrane potential near the channel. Figure 3.45 shows the X-ray crystal structure of bacterial KcsA potassium ion channel, in which four subunits form a channel and potassium ions are in the channel. The membrane potential changes the conformation of the channel proteins and regulates their opening and closing. The transmembrane protein channels play an important role in excitable cells such as neurons and muscle tissue, enabling rapid and coordinated depolarization in response to the triggering of voltage changes.

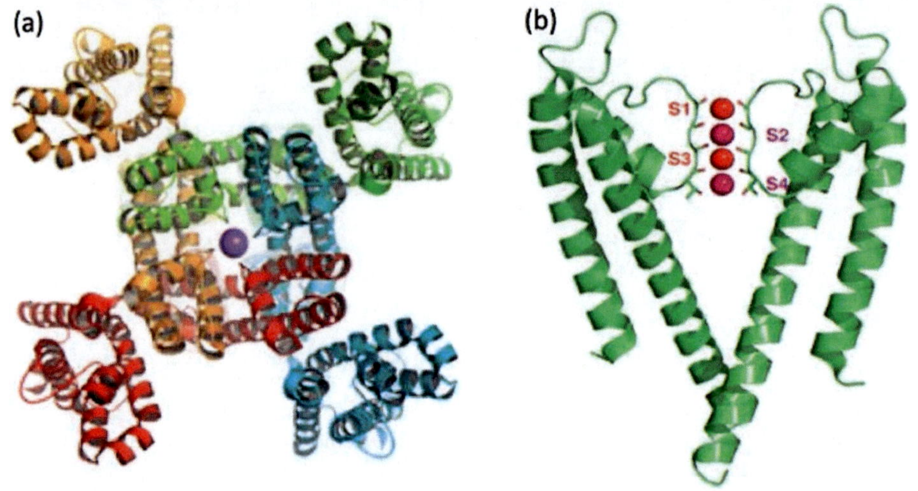

Fig. 3.45 Structure of voltage-gated potassium channel (pdb:2R9R). **a** The X-ray crystal structure of bacterial KcsA potassium ion channel, in which four subunits form a channel and potassium ions are in the channel. **b** Two of the four subunits are shown for clarity. The potassium ions occupying the S2 and S4 sites and the oxygen atoms of water molecules in S1 and S3 sites are depicted as purple and red spheres, respectively

3.5.2.1 K^+ Channel Block with TEA

Tetraethylammonium (TEA) has long been known as a blocker of open-type ion channels. The cocrystal structure of TEA with the potassium ion selective channel KcsA of the bacterium Streptomyces lividans (2BOC) shows that TEA is located at the entrance of the KcsA channel (Fig. 3.46) [32]. Four aromatic rings of Tyr82 located at the top of the tetrameric channel surround TEA. Ab initio calculations for the orientation between the benzene rings of Tyr82 and TEA based on the X-ray crystal structure show that an *en face* geometry is predominant, suggesting that the benzene rings are thought to be involved in TEA recognition through cation-π interactions [33].

In order to clarify the recognition of TEA by aromatic residues, Ahern et al. used KcsA incorporated a series of fluorinated Phe side chains at position 449 and investigated the degree of inhibition by TEA [33]. As shown in Table 3.12, fluorination of the aromatic ring resulted in a gradual increase in the inhibition constant K_i for the TEA block. A larger increase in K_i was observed when the benzene ring of Phe449 was replaced with a non-aromatic cyclohexane (X=Cha). These results support the contribution of a strong cation-π interaction in the TEA block.

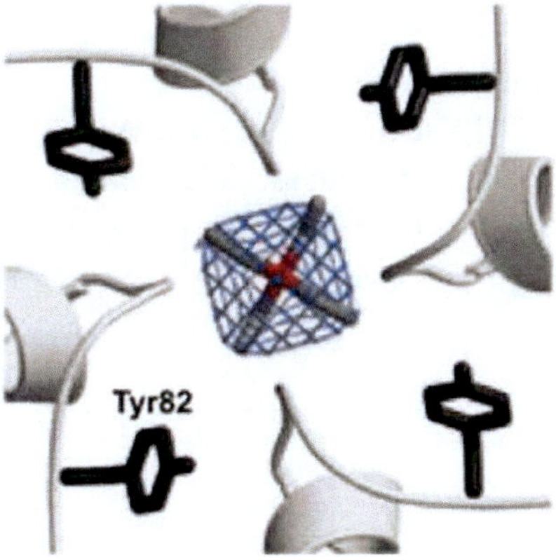

Fig. 3.46 View down the fourfold axis onto the TEAs-binding site. The selectivity filter and part of the pore helix are gray, and the side chains of Tyr82 are highlighted. The TEA omit map contoured at 2.5 σ is shown as a blue mesh. TEAs are within van der Waals range of the Tyr82 side chains: the TEAs(C2)-Tyr(CΣ) distances are ~4.1 Å on all subunits. Reprinted by permission from Springer Nature [32], Copyright Springer Nature Singapore Pte Ltd. 2005

Table 3.12 Inhibitory constants for TEA block based on fits of dose–response curves

T449X, where X =	K_i(mM)	$\Delta\Delta G$(kcal/mol)
Phe	0.39±0.04	0
Thr	33.5±8.9	2.58
4F-Phe	3.3±0.6	1.23
3,5F-Phe	47.6±5.9	2.78
3,4,5F-Phe	175.2±36.2	3.54
Cha	249.2±20.1	3.75

[a] K_i values are given as the mean+SEM. [b] Free energy ($\Delta\Delta G$) is calculated as $\Delta\Delta G = RT\ln(K_i, X/K_i, Phe)$, where RT is 0.58 kcal/mol at room temperature

3.5.2.2 Na^+ Channel Block with TTX

Tetrodotoxin (TTX) is known as an inhibitor of voltage-dependent sodium channels; TTX blocks sodium channels with nanomolar affinity only when aromatic Phe or Tyr residues are present at specific positions in the external vestibule of the ionic conduction pore. Ahern et al. focused on Tyr401 in rat skeletal muscle sodium channels (NaV1.4) with respect to inhibition by TTX [34]. To clarify the role of Tyr401, this aromatic residue was replaced with a fluorinated derivative of Phe. Increased fluorination of Phe401 decreases the negative electrostatic potential of the aromatic face, which caused an increase in the blocking inhibition constant. Trifluorination of the aromatic ring resulted in a 50-fold decrease in TTX affinity. This decrease is similar to that of the substitution with hydrophobic Leu. The electrostatic potential of TTX shown in Fig. 3.47 clearly shows that the guanidinium moiety is positively charged and blocks the ion channel due to the cation-π interaction between the guanidinium moiety and Phe401.

The binding model of TTX to the benzene ring was optimized by QM/MM calculations. As shown in Fig. 3.48, the aromatic side chain of Tyr401 faces the permeation pathway, and the guanidinium ion part of TTX approaches Tyr401 by a cation-π interaction, blocking the permeated ion.

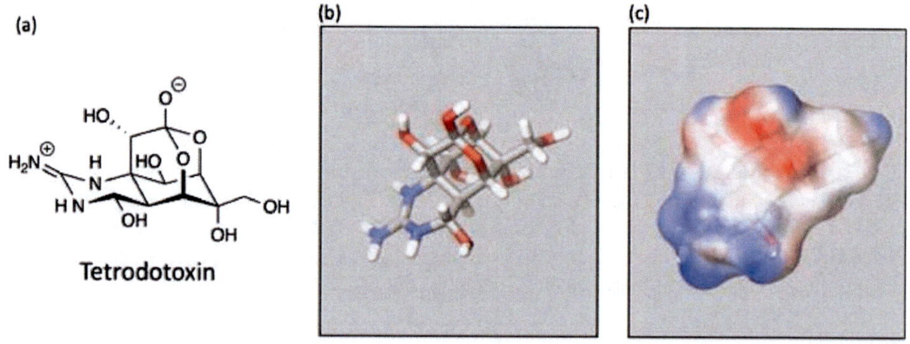

Fig. 3.47 **a** Molecular structure of TTX. **b** Energy-minimized structure. **c** Electrostatic potential ranging from +100 kcal/mol (blue) to +40 kcal/mol (red). Reprinted from [34], Copyright 2007, with permission from Elsevier

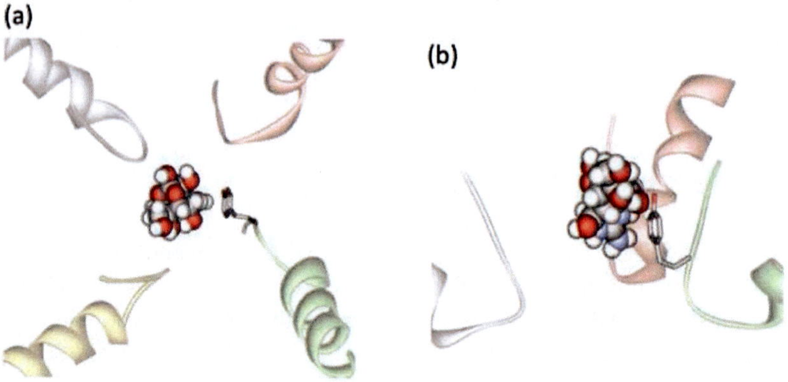

Fig. 3.48 Model of TTX docking in the pore mouth of Nav1.4. **a** Top view of TTX interacting with the pore loops of domains DI-DIV in the Lipkind-Fozzard model. This structure represents one possible configuration of the encounter complex. **b** Side view of TTX in the pore. Reprinted from [34], Copyright 2007, with permission from Elsevier

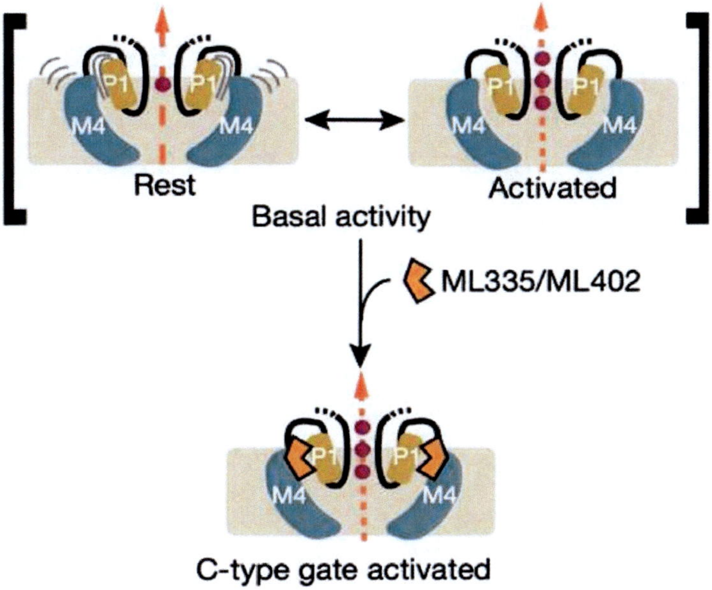

Fig. 3.49 TREK activation model. Gray lines indicate mobile P1 (tan) and M4 (blue). C-type activators (orange) stabilize the selectivity filter and channel "leak mode." Potassium ions are purple. Gap in arrows indicates current flow intensity. Membrane is gray (Reproduced from Lolicato et al. 2017) [35]

3.5.2.3 K_{2p} Channel Activation

TREK-1 is the best-studied two-pore domain potassium channel (K_{2p}) and, like other potassium ion channels, has a tetrameric structure. Its primary role is to regulate cell excitability and maintain membrane potential below the threshold of depolarization. TREK-1 has been implicated in many physiological and pathological processes in the nervous system, including depression, neuroprotection, pain, and anesthesia.

In contrast to other potassium channels, the K_{2P} channel uses a "C-type" gate with a selective filter as its primary gating site. As shown in Fig. 3.49, P1 and M4 are fixed in position and the C-type gate is stimulated by ligand uptake. The small molecules, N-arylsulfonamide ML335 and thiophenecarboxamide ML402 (Fig. 3.50), are known to act as TREK activators. Figure 3.49 shows that they directly stimulate the C-type gate by binding to TREK [35]. ML335 and ML402 have an L-shaped conformation, with aromatic ring interactions with the Phe134 residue of P1 helix and the Trp275 residue of M4, and cation-π interactions with the Lys271 residue of M4. In addition, there are hydrogen bonds between the amide groups and Lys271 (Fig. 3.51).

These interactions fix the movement of the P1 and M4 sites, making the ion filter to keep in leak mode, which activates potassium ion permeability. Since the basic

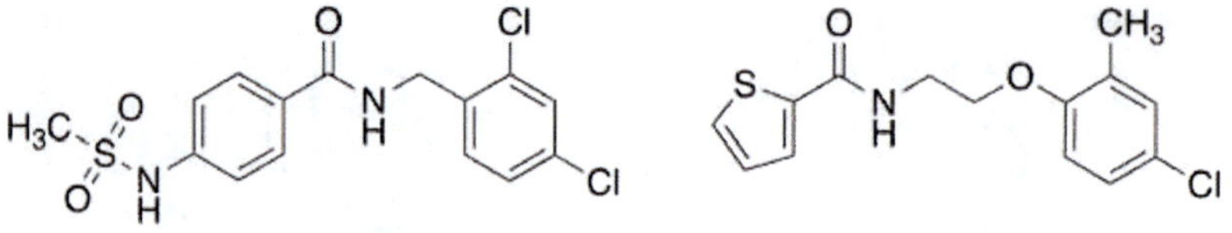

ML335 ML402

Fig. 3.50 Molecular structure of ML335 and ML402

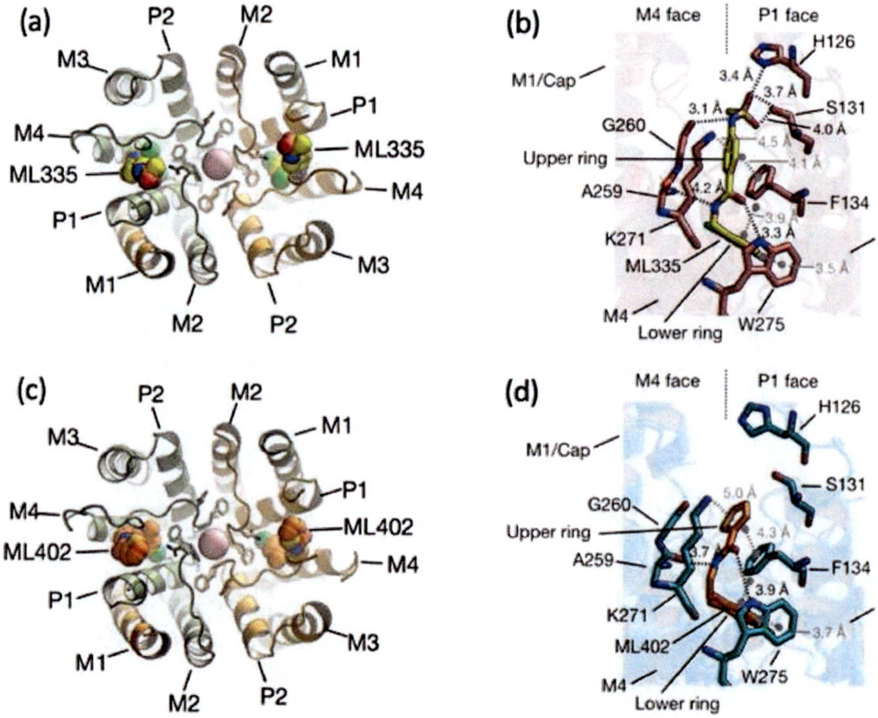

Fig. 3.51 Extracellular views excluding the cap domain of ML335 **a** and ML402 **c** binding sites. ML335 and ML402 are space filling, and selectivity filter side chains are sticks. Cartoon diagram of K2P2.1–ML335 **b** and K2P2.1–ML402 **d** interactions. Reprinted by permission from Springer Nature [35], Copyright Springer Nature Singapore Pte Ltd. 2017

pocket structure is conserved in the voltage-gated ion channel (VGIC) superfamily, similar regulatory mechanisms may exist in other superfamilies where selectivity filter-based gating is central.

3.5.2.4 M2 Proton Channel Activation

The M2 protein of influenza viruses forms a homotetrameric proton channel that conducts protons to the virion under acidic pH. The proton selectivity of this tetrameric channel is regulated by one histidine (His37), and channel gating is achieved by one tryptophan (Trp41) in the transmembrane domain of the protein.

To elucidate the cation-π interaction between these two functional residues, Hong et al. used high-resolution solid-state NMR spectroscopy to determine the side chain conformation and dynamics of Trp41 in the M2 transmembrane peptide [36]. Figure 3.52 gives examples of the reported Trp41 rotamers in the M2 protein, with four tryptophans forming the gate of the channel. Figure 3.53 shows the equilibrium rotamer conformation of Trp and its dynamic changes at high and low pH. To show the extent to which the indole ring fluctuates, two marginal rotamers are shown, one standard deviation (σ) away from the equilibrium χ_1 and χ_2 angles. These results clearly show that at acidic pH the average position of the indole ring is closer to the imidazolium than at high pH.

To give a quantitative estimation of the His-Trp proximity, the distances from His37 N ε 2 to the centers of the benzene and pyrrole rings were calculated. The distances between His and the benzene and pyrrole rings in the equilibrium Trp41 conformation at low pH were 4.3 Å and 3.0 Å, respectively, whereas at high pH they are 5.4 Å and 3.7 Å, respectively. These experimental and computational results suggest that the cation-π interaction between protonated His37 and Trp41 significantly contributes to the dynamics of Trp41. The contact between His37 and Trp41 at low pH may be an important reason for the dynamic controlling of the proton flux in the M2 channel.

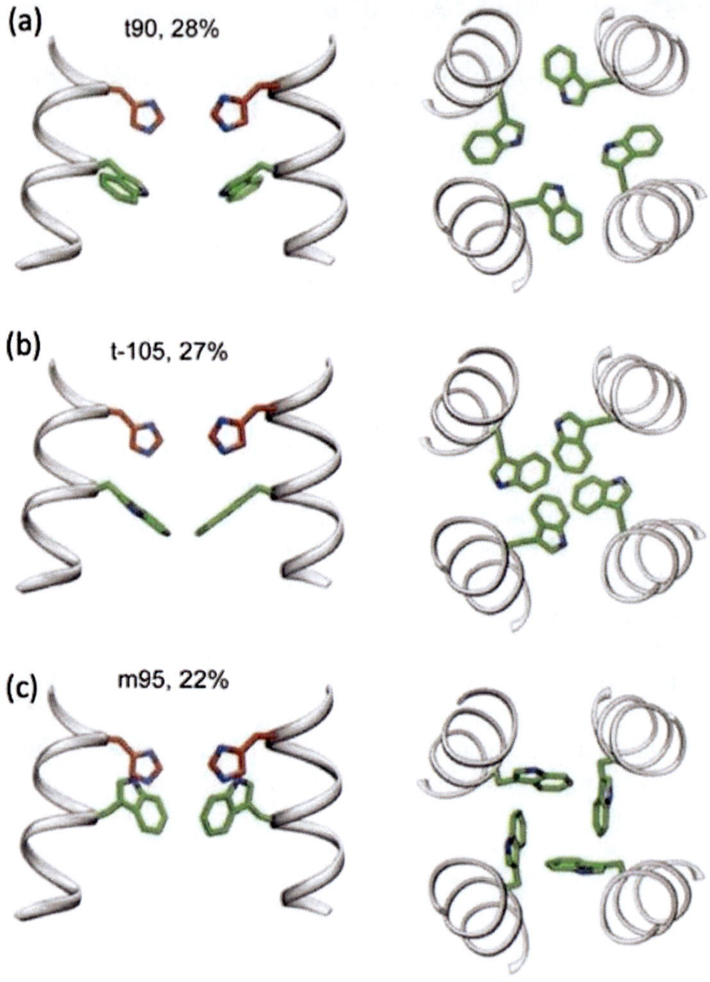

Fig. 3.52 Several reported Trp41 rotamers using various experimental techniques. (Left column) Side view of two of the four helices; both His37 and Trp41 side chains are depicted. (Right column) Top view of the four helices and the Trp41 side chains. **a** t90 rotamer ($\chi 1 = 180^\circ$, $\chi 2 = +90^\circ$). **b** t-105 rotamer ($\chi 1 = 180^\circ$, $\chi 2 = -105^\circ$). **c** m95 rotamer ($\chi 1 = -60^\circ$, $\chi 2 = +95^\circ$). Reprinted from [36], Copyright 2013, with permission from Elsevier

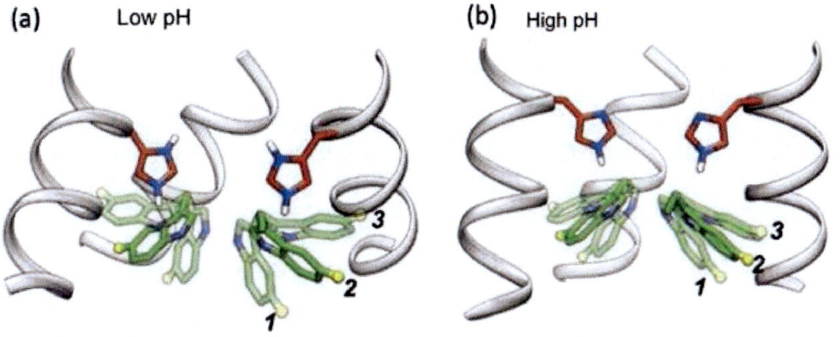

Fig. 3.53 Trp41 conformational dynamics and aromatic interaction with His37. **a** Low-pH scenario. The measured equilibrium ($\chi 1$, $\chi 2$) rotamer of (155°, 135°) is shown as position 2, together with two limiting rotamers based on the measured order parameters. The (125°, 150°) rotamer (position 3) is closer to His37, whereas the (175°, 120°) rotamer (position 1) is further away from His37. These angles were chosen to be one standard deviation (30° for $\sigma 1$ and 15° for $\sigma 2$) from the average angle. **b** High-pH scenario. The equilibrium rotamer is (175°, 120°) (position 2), and the limiting rotamers are (150°, 120°) (position 1) and (160°, 120°) (position 3), based on a $\sigma 1$ of 25° and no $\chi 2$ fluctuation. Reprinted from [36], Copyright 2013, with permission from Elsevier

References

1. Gallivan JP, Dougherty DA (1999) Cation-π interactions in structural biology. Proc Natl Acad Sci USA 96:9459–9464
2. Shi Z, Olson CA, Kallenbach NR (2002) Cation-π interaction in model α-helical peptides. J Am Chem Soc 124:3284–3291
3. Tsou LK, Tatko CD, Waters ML (2002) Simple cation-π interaction between a phenyl ring and a protonated amine stabilizes an α-helix in water. J Am Chem Soc 124:14917–14921
4. Tatko CD, Waters ML (2003) The geometry and efficacy of cation-π interactions in a diagonal position of a designed β-hairpin. Protein Sci 12(11):2443–2452
5. Hughes RM, Waters ML (2005) Influence of N-methylation on a cation-π interaction produces a remarkably stable β-hairpin peptide. J Am Chem Soc 127:6518–6519
6. Riemen AJ, Waters ML (2009) Design of highly stabilized β-hairpin peptides through cation-π interactions of lysine and N-Methyllysine with an aromatic pocket. Biochem 48:1525–1531
7. Craven TW, Cho MK, Traaseth NJ, Bonneau R, Kirshenbaum K (2016) A miniature protein stabilized by a cation-π interaction network. J Am Chem Soc 138:1543–1550
8. Goldstein R, Cheng J, Stec B, Roberts MF (2012) Structure of the S. aureus-π-specific phospholipase C reveals modulation of active site access by a titratable-π-cation latched loop. Biochemistry 51:2579–2587
9. Crowley PB, Golovin A (2005) Cation-π interactions in protein-protein interfaces. Proteins 59:231–239
10. Johnson RM, Heslop CL, Deber CM (2004) Hydrophobic helical hairpins: design and packing interactions in membrane environments. Biochem 43:14361–14369
11. Johnson RM, Hecht K, Deber CM (2007) Aromatic and cation-π interactions enhance helix-helix association in a membrane environment. Biochem 46:9208–9214
12. Chen CC, Hsu W, Kao TC, Horng JC (2011) Self-assembly of short collagen-related peptides into fibrils via cation-π interactions. Biochem 50:2381–2383
13. Chiang CH, Horng JC (2016) Cation-π interaction induced folding of AAB-type collagen heterotrimers. J Phys Chem B 120:1205–1211

14. Chiang CH, Fu YH, Horng JC (2017) Formation of AAB-type collagen heterotrimers from designed cationic and aromatic collagen-mimetic peptides: evaluation of the C-terminal cation-π interactions. Biomacromol 18:985–993
15. Wu D, Hu Q, Yan Z, Chen W, Yan C, Huang X, Zhang J, Yang P, Deng H, Wang J, Deng XW, Shi Y (2012) Structural basis of ultraviolet-B perception by UVR8. Nature 484:214–219
16. Christie JM, Arvai AS, Baxter KJ, Heilmann M, Pratt AJ, O'Hara A et al (2012) Plant UVR8 photoreceptor senses UV-B by tryptophan-mediated disruption of cross-dimer salt bridges. Science 335:1492–1496
17. Gromiha MM, Santhosh C, Suwa M (2004) Influence of cation–π interactions in protein–DNA complexes. Polym 45:633–639
18. Worch R, Jankowska-Anyszka M, Niedzwiecka A, Stepinski J, Mazza C, Darzynkiewicz E et al (2009) Diverse role of three tyrosines in binding of the RNA 5' cap to the human nuclear cap binding complex. J Mol Biol 385(2):618–627
19. Liang Z, Li QX (2018) π-Cation interactions in molecular recognition: perspectives on pharmaceuticals and pesticides. J Agric Food Chem 66(13):3315–3323
20. Schärer K, Morgenthaler M, Paulini R, Obst-Sander U, Banner DW, Schlatter D et al (2005) Quantification of cation-π interactions in protein-ligand complexes: crystal-structure analysis of factor Xa bound to a quaternary ammonium ion ligand. Angew Chem Int Ed Engl 44(28):4400–4404
21. Salonen LM, Bucher C, Banner DW, Haap W, Mary JL, Benz J et al (2009) Cation-π interactions at the active site of factor Xa: dramatic enhancement upon stepwise N-alkylation of ammonium ions. Angew Chem Int Ed Engl 48(4):811–814
22. Faraldos JA, Antonczak AK, González V, Fullerton R, Tippmann EM, Allemann RK (2011) Probing eudesmane cation-π interactions in catalysis by aristolochene synthase with non-canonical amino acids. J Am Chem Soc 133:13906–13909
23. Faraldos JA, Allemann RK (2011) Inhibition of (+)-aristolochene synthase with iminium salts resembling eudesmane cation. Org Lett 13:1202–1205
24. Zheng P, Xiang L, Chang J, Lin Q, Xie L, Lan T et al (2021) Nanomechanics of lignin-cellulase interactions in aqueous solutions. Biomacromol 22:2033–2042
25. Sureshan KM, Trusselle M, Tovey SC, Taylor CW, Potter BVL (2008) 2-Position base-modified analogues of adenophostin a as high-affinity agonists of the D-myo-inositol trisphosphate receptor: in vitro evaluation and molecular modeling. J Org Chem 73:1682–1692
26. Morales-Perez CL, Noviello CM, Hibbs RE (2016) X-ray structure of the human α4β2 nicotinic receptor. Nature 538(7625):411–415
27. Van Arnam EB, Dougherty DA (2014) Functional probes of drug-receptor interactions implicated by structural studies: Cys-loop receptors provide a fertile testing ground. J Med Chem 57(15):6289–6300
28. Xiu X, Puskar NL, Shanata JA, Lester HA, Dougherty DA (2009) Nicotine binding to brain receptors requires a strong cation-π interaction. Nature 458(7237):534–537
29. Blom AEM, Campello HR, Lester HA, Gallagher T, Dougherty DA (2019) Probing binding interactions of cytisine derivatives to the α4β2 nicotinic acetylcholine receptor. J Am Chem Soc 141(40):15840–15849
30. Duffy NH, Lester HA, Dougherty DA (2012) Ondansetron and granisetron binding orientation in the 5-HT(3) receptor determined by unnatural amino acid mutagenesis. ACS Chem Biol 7:1738–1745
31. Lummis SC, McGonigle I, Ashby JA, Dougherty DA (2011) Two amino acid residues contribute to a cation-π binding interaction in the binding site of an insect GABA receptor. J Neurosci 31(34):12371–12376
32. Lenaeus MJ, Vamvouka M, Focia PJ, Gross A (2005) Structural basis of TEA blockade in a model potassium channel. Nat Struct Mol Biol 12:454–459
33. Ahern CA, Eastwood AL, Lester HA, Dougherty DA, Horn R (2006) A cation-π interaction between extracellular TEA and an aromatic residue in potassium channels. J Gen Physiol 128(6):649–657

34. Santarelli VP, Eastwood AL, Dougherty DA, Horn R, Ahern CA (2007) A cation-π interaction discriminates among sodium channels that are either sensitive or resistant to tetrodotoxin block. J Biol Chem 282(11):8044–8051
35. Lolicato M, Arrigoni C, Mori T, Sekioka Y, Bryant C, Clark KA, Minor Jr D (2017) K_{2P}2.1 (TREK-1)-activator complexes reveal a cryptic selectivity filter binding site. Nature 547(7663):364–8
36. Williams JK, Zhang Y, Schmidt-Rohr K, Hong M (2013) pH-Dependent conformation, dynamics, and aromatic interaction of the gating tryptophan residue of the influenza M2 proton channel from solid-state NMR. Biophys J 104(8):1698–1708

Chapter 4
Organic Synthesis

4.1 Role of Cation–π Interactions in Organic Reactions

The large interaction energies of cation–π interactions and the diversity of combinations of cations and π-electron systems are described in Chap. 2. In addition, cation–π interactions between the substrate and the enzyme receptor play a key role in enzymatic reactions, as discussed in Chap. 3. These features indicate that cation–π interactions have tremendous potential in terms of controlling organic reactions [1].

Cation–π interactions are thought to contribute to the stabilization of the cationic substrates, products, intermediates, or transition states of organic reactions. Figure 4.1a–d show schematic representations of the participation of a π-component to the cationic species, before reaction, in transition state, in intermediate, and after reaction, respectively. When cationic substrate **A** is in equilibrium between the two conformers, **A** and **A'**, and conformer **A'** is stabilized by a π-component, the reaction should occur from conformer **A'** (Fig. 4.1a). Figure 4.1b shows that stabilization of the cationic transition state of the reaction by the interaction with a π-component allows the activation energy to be lowered, thus increasing the reaction rate of the desired reaction. Figure 4.1c shows the reaction via two cation intermediates **B** and **C**, where intermediate **C** is stabilized by a π-component and the reaction undergoes from **C**. Figure 4.1d shows that when cationic products **B** and **C** are in equilibrium, **C** can be obtained as the major product by stabilizing **C** with a π-component. If the substrate, transition state, intermediates, and products are neutral, a similar explanation is possible by replacing the π-component in Fig. 4.1 with a cation. An appropriate use of cation–π interactions in these stages allows control of the product selectivity.

S. Yamada, *The Cation-π Interaction*,
https://doi.org/10.1007/978-981-19-7335-2_4

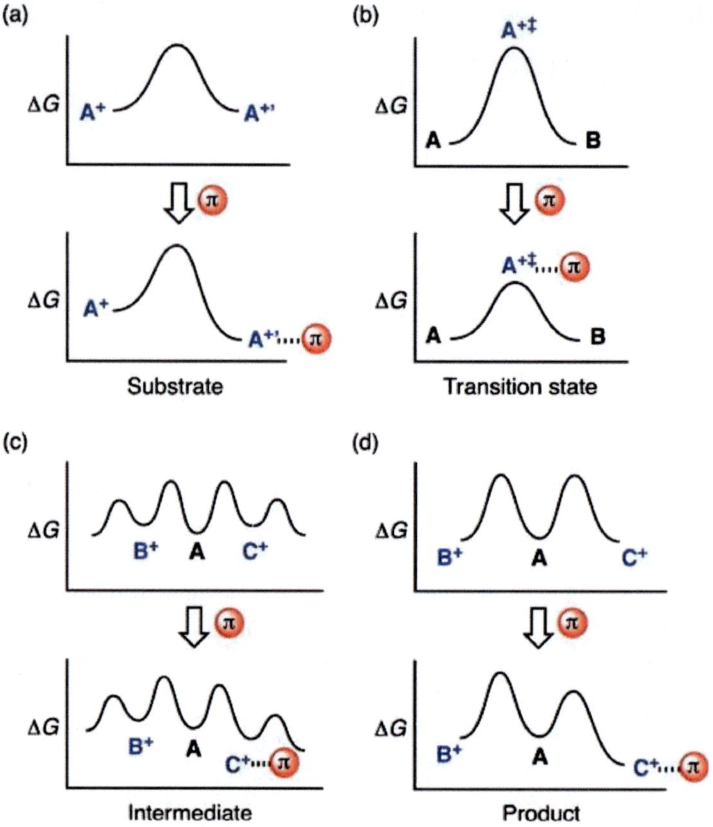

Fig. 4.1 Schematic representation for the participation of a cation with a molecule during a reaction: interaction with the substrate **a** before the reaction, **b** in the transition state, **c** in the intermediate, and **d** interaction with the product after the reaction

4.2 Organic Cations

There are various organic cations, including pyridinium, iminium, oxonium, and carbonium ions, and the reactivity of these cations depends on their structures. Stable cations can be used to assist and control reactions through interaction with substrates as explained in Fig. 4.1. On the other hand, unstable cations are generally observed in the reaction intermediates and are employed for various synthetic reactions. The intra- and intermolecular cation–π interactions involving these cations are effective tools for controlling various types of reactions.

4.2.1 *Pyridinium and Heteroaromatic Cations*

4.2.1.1 Dihydropyridine Synthesis

Asymmetric synthesis of chiral 1,4-dihydropyridines has received considerable attention because of their synthetic utility for various natural products and bioactive compounds. One effective method for the synthesis of chiral 1,4-dihydropyridines is the face-selective addition of nucleophiles to pyridinium salts. Yamada et al. used nicotinamides **1a** and **1b** possessing chiral oxazolidine moieties [2]. The pyridinium intermediates can be generated by the reaction of **1a** and **1b** with methyl chloroformate, and the addition of ketene silyl acetals to the intermediates afforded 1, 4-dihydropyridines **2a** and **2b** in 12% and 99% de, respectively (Scheme 4.1a). This large difference in diastereoselectivity is thought to be due to differences in the conformation of the pyridinium intermediates.

X-ray crystallographic analysis of **1b** and *N*-methylpyridinium salt **4** as a model for the *N*-acylpyridinium intermediate clarified the differences in their geometries. As shown in Fig. 4.2, the phenyl and pyridinium rings are placed in close proximity and face each other, with the distance between the two rings determined to be about 3.4 Å. This structural difference between them clearly shows that a cation–π interaction is effective in controlling the conformation of the pyridinium intermediate. CD spectral studies and the NMR experiments on **4** suggest that a closed conformation is preferred, even in solution. DFT calculations indicate that the equilibrium between **3b**-I and **3b**-II favors **3b**-II. This supports the proposed mechanism whereby the nucleophile attacks conformer II from the less hindered side to give the chiral dihydropyridine **2b** (Scheme 4.1b).

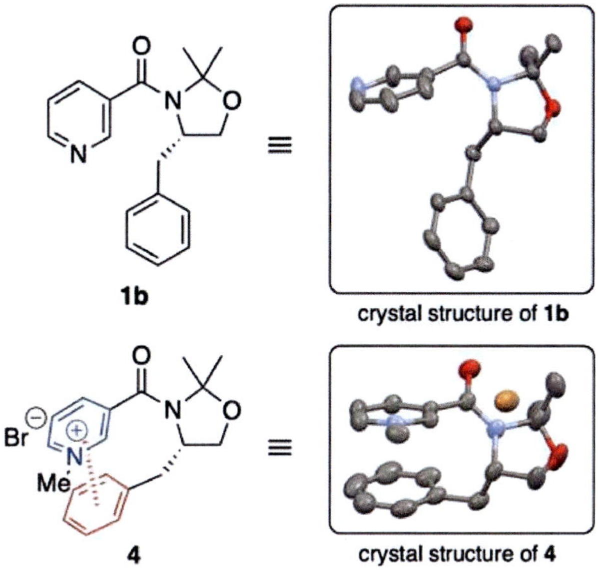

Fig. 4.2 X-ray crystal structures of **1b** and **4**. Reprinted with permission from [1]. Copyright 2018 American Chemical Society

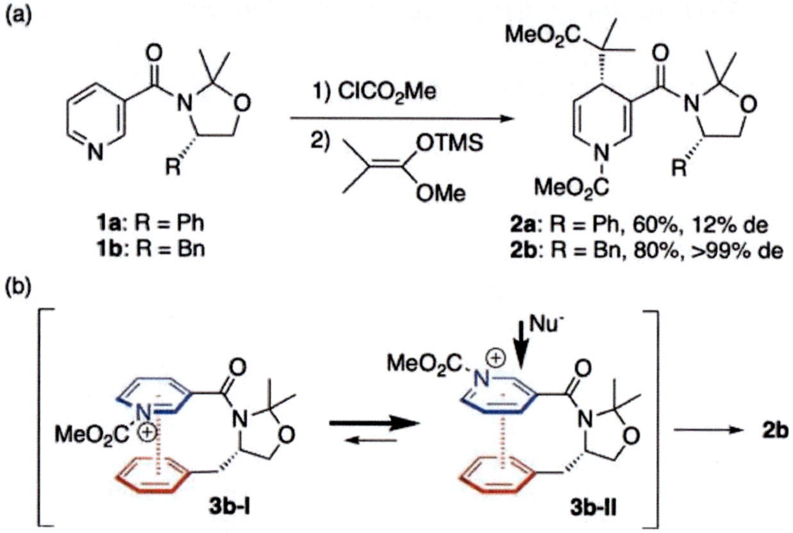

Scheme 4.1 Face-selective addition of a ketene silyl acetal to a chiral pyridinium salt. Reprinted with permission from [1]. Copyright 2018 American Chemical Society

4.2.1.2 Dihydropyridone Synthesis

Comins et al. developed a general synthetic procedure for the synthesis of chiral dihydropyridones using chiral chloroformate with a (-)-8-phenylmenthyl (8-PhM) moiety, as shown in Scheme 4.2 [3]. Addition of $Cl(CH_2)_4MgBr$ to acylpyridinium salt **5** prepared in situ from 4-methoxy-3-(triisopropylsilyl)pyridine and chiral chloroformate produced 6-substituted dihydropyridone **6** in 77% yield with 86% de (Scheme 4.2). The major diastereomer of **6** has an *R* configuration at C6 of the dihydropyridone ring. This good diastereoselectivity can be explained by the stabilization of TS-6 through cation–π interactions between the pyridinium and the aromatic ring of the chiral auxiliary, as shown in Scheme 4.2. This was used for the synthesis of the quinolizidine alkaloids (+)-myrtin, (-)-rasvin, and (+)-subcosin I. The X-ray crystal structure of model compound **7** shows the two rings to be in close proximity (Fig. 4.3) [4], supporting the transition state model TS-6 (Scheme 4.2).

4.2.1.3 Catalytic Acylation and Deacylation

Yamada et al. developed a chiral DMAP catalyst **8** bearing a chiral thiazolidine-2-thione moiety at the C3 of dimethylaminopyridine by an amide bond [5, 6]. This asymmetric acylation catalyst was effective in the kinetic resolution of various racemic *sec*-alcohols [5], desymmetrization of *meso*-diols [5], and dynamic

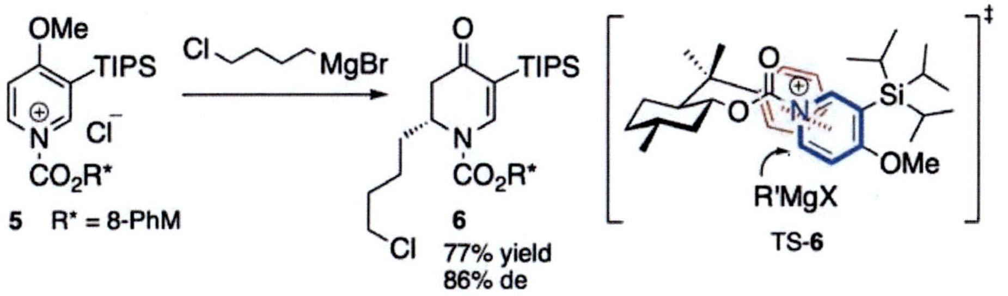

Scheme 4.2 Stereoselective addition of a Grignard reagent to a pyridinium salt

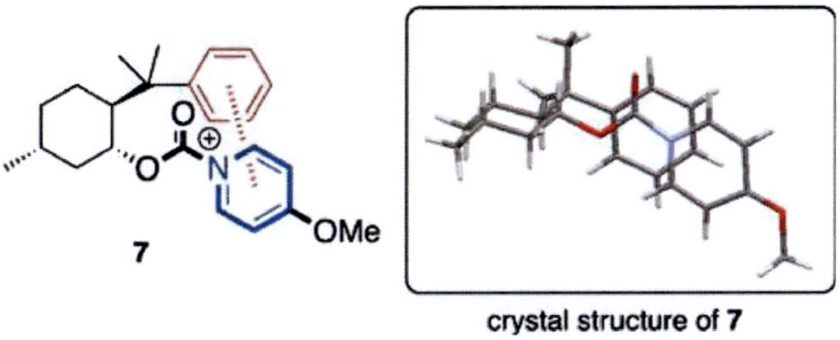

Fig. 4.3 Crystal structure of compound **7**. Reprinted with permission from [1]. Copyright 2018 American Chemical Society

kinetic resolution of hemiaminals [6]. (Scheme 4.3). The conformation of the *N*-acylpyridinium intermediate **9** is fixed by a cation–π interaction between the pyridinium ring and the N–C=S moiety, as shown in Fig. 4.4, leading to a chiral environment around the pyridinium. A comparison of the X-ray crystal structures of the model compound **16** and the corresponding N-methyl derivative **17** clearly shows that the thiocarbonyl of **17** is closer to the pyridinium ring than is that of **16** (Fig. 4.4b). ^{1}H NMR spectroscopic studies and DFT calculations of the N-acyl intermediate at the B3LYP/6-31G (d) level confirmed that the geometry is close to that of the crystal structure of **17**. In addition to the intramolecular cation–π interaction, the intermolecular cation–π interaction between the pyridinium intermediate and the aromatic ring of the substrate is thought to be key to the resolution of racemic alcohols by stabilization of the transition state TS-**10**-i (Fig. 4.4a).

Birman and coworkers developed an amidine-based enantiomer-selective acyl-transfer catalyst **18** [7]. The kinetic resolution of various secondary benzyl alcohols using this catalyst was successful in providing chiral alcohols in a highly selective manner. For example, the asymmetric acylation of 1-phenylpropyl alcohol **19** with catalyst **18** yielded ester **20** and recovered chiral alcohol **19** in high selectivity ($s = 36$) (Scheme 4.4a). The lowest-energy transition state TS-20 obtained by theoretical calculations at the B3LYP/6-31G(d) level indicates that the aryl group of the substrate stacks on the pyridinium ring of N-acyl intermediate **21** through cation–π interactions, which can explain the observed enantioselectivity (Scheme 4.4b).

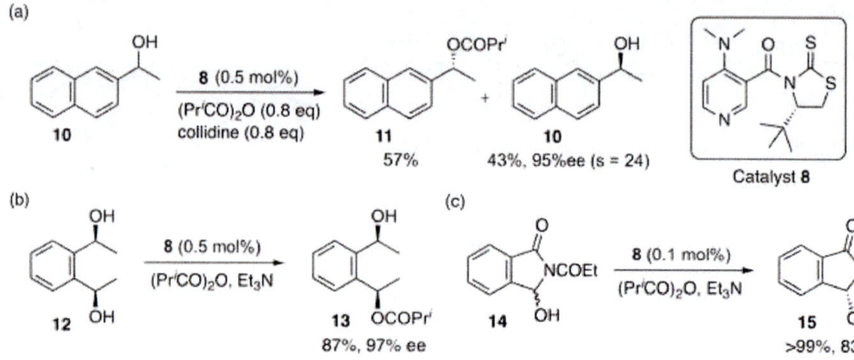

Scheme 4.3 Asymmetric acylation of *sec*-alcohols using chiral DMAP catalyst **8**: **a** Kinetic resolution of racemic *sec*-alcohols, **b** desymmetrization of *meso*-diol, and **c** dynamic kinetic resolution of hemiaminals

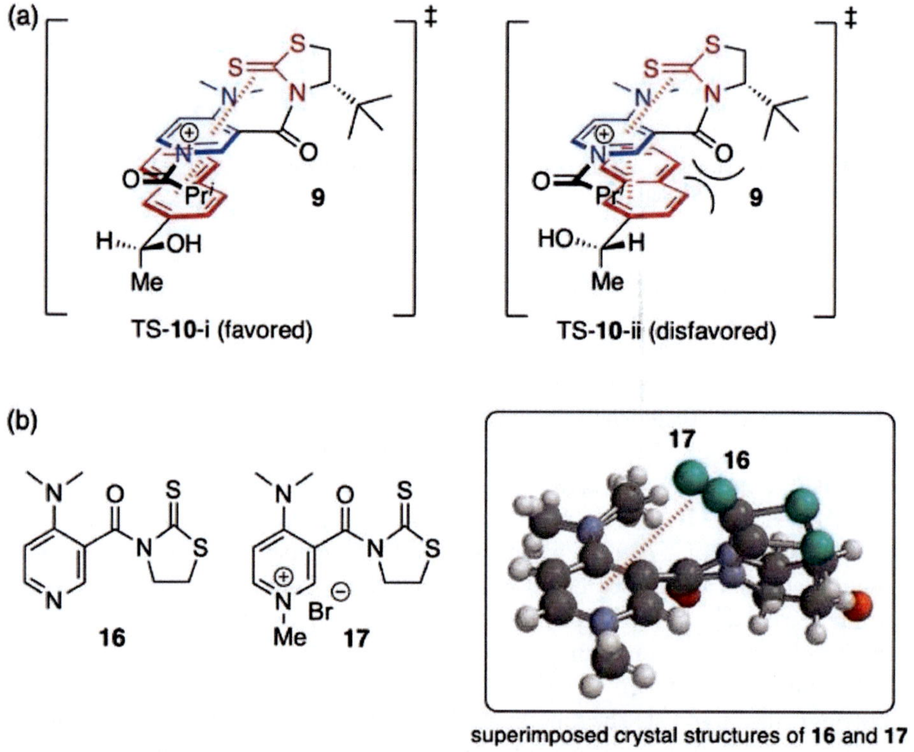

Fig. 4.4 **a** Hypothesized transition structures of the kinetic resolution and **b** superimposed crystal structures of **16** and **17**. Reprinted with permission from [1]. Copyright 2018 American Chemical Society

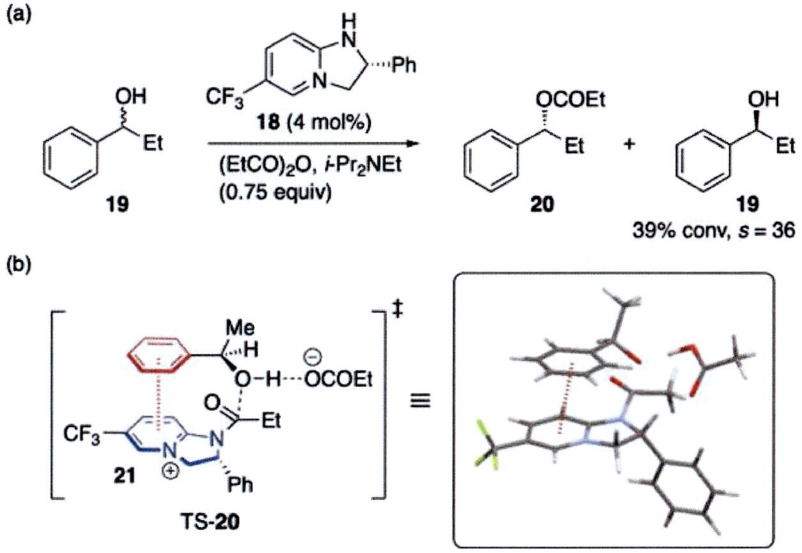

Scheme 4.4 Kinetic resolution of racemic *sec*-alcohol **19** with the amidine-based catalyst **18**. Reprinted with permission from [1]. Copyright 2018 American Chemical Society

Waters et al. developed α-hairpin peptide catalyst for the deacylation of esters [8]. As shown in Scheme 4.5, the 8-mer peptide **23** catalyzed the deacylation of nitrophenyl methoxyacetate **22** to give ester **24** and *p*-nitrophenol **25**. The rate of the reaction is accelerated when the R substituent is an aromatic ring, and a methoxy substituent at the benzene ring is effective in accelerating the ester exchange reaction. NOE experiments revealed that the histidine protons are in close proximity to the benzene ring, indicating that the cation–π interaction influences the folding suitable for the deacylation reaction.

4.2.1.4 Steglich Reaction

Smith et al. reported the catalytic enantioselective [2,3] rearrangement of allylammonium ylides catalyzed by isothiourea [9]. Ammonium salt **26** was rearranged in the presence of iPr$_2$NH, BnNH$_2$, and HOBt at −20 °C with 20 mol% of benzotetramisole (BTM) **28** to give [2,3]-rearranged amino acid derivative **27** in a 95:5 dr and 99% ee (Scheme 4.6). In this reaction, catalyst **28** serves as a nucleophilic catalyst and [2,3] rearrangement successively occurs via the imido intermediate **29**. The calculations of this transition state show TS-27 as the lowest-energy transition structure, in which the catalyst and the aromatic ring of the substrate are close together, suggesting the contribution of a cation–π interaction to the high selectivity.

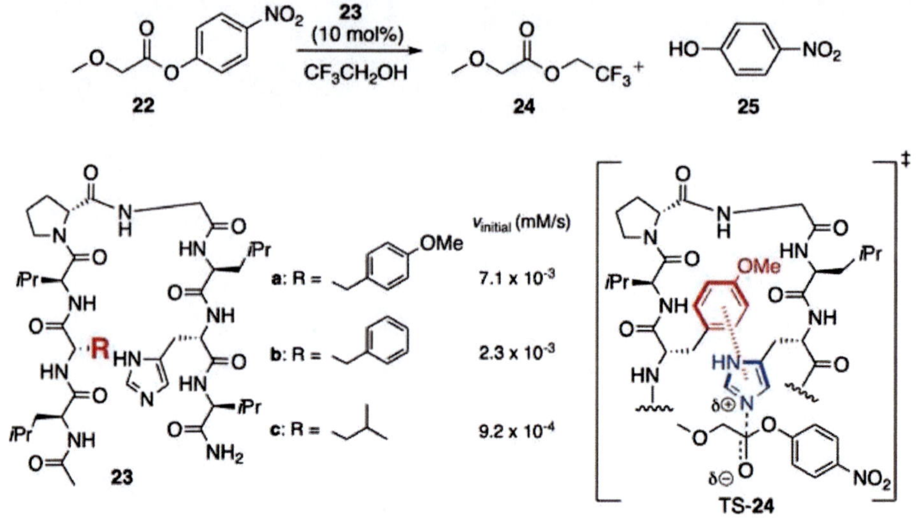

Scheme 4.5 Peptide-catalyzed deacylation reactions of ester **22**. Reprinted with permission from [1]. Copyright 2018 American Chemical Society

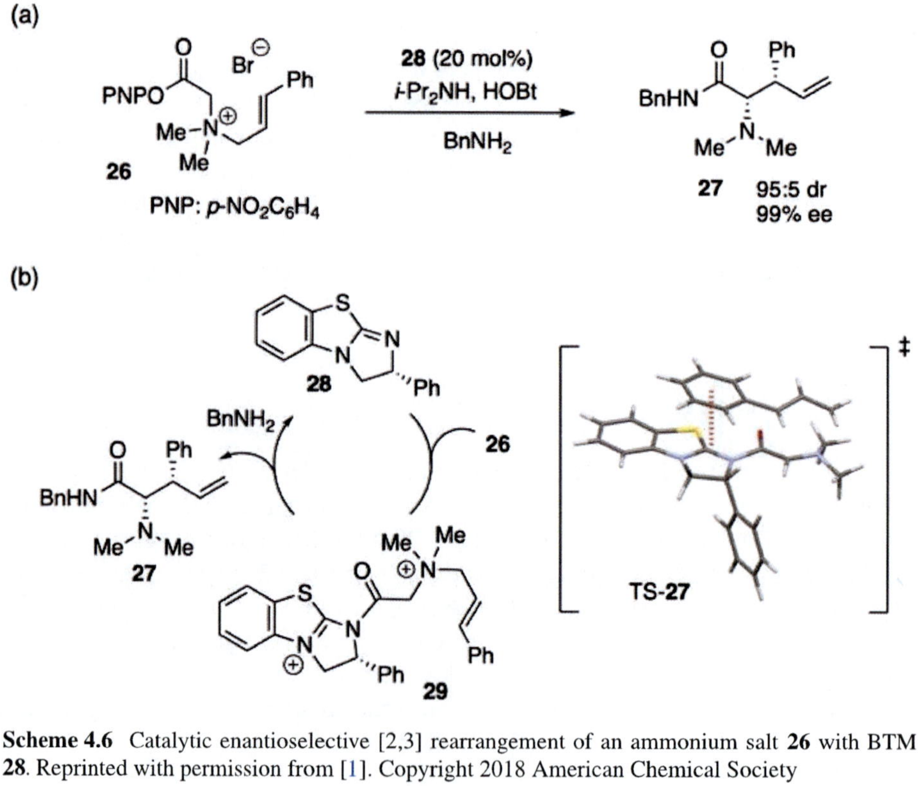

Scheme 4.6 Catalytic enantioselective [2,3] rearrangement of an ammonium salt **26** with BTM **28**. Reprinted with permission from [1]. Copyright 2018 American Chemical Society

4.2.1.5 [2+2] and [4+4] Photodimerization

[2+2] and [4+4] photocyclizations are fundamental and important reactions in organic photochemistry and have been applied to the synthesis of various natural products. In general, these dimerization reactions produce many isomers; therefore, preorientation of substrates in solution is an important issue. Williams et al. conducted pioneering work on the [2+2] photodimerization of styrylpyridines [10]. They found that irradiation of 2-styrylpyridine **30** in solution produced only a mixture of (*Z*)-and (*E*)-2-styrylpyridines, whereas irradiation of 4-styrylpyridine in dilute hydrochloric acid yields dimers.

Yamada et al. presumed that cation–π interactions play a major role in the yield and stereoselectivity of photodimer formation and systematically investigated the effects of the substituent and hydrochloric acid concentration on the photodimerization reaction of 4-substituted (*E*)-4-styrylpyridines [11]. Irradiation of (*E*)-4-styrylpyridine **30a** in methanol under neutral conditions afforded (*Z*)-**30a** as the major product with three dimers **31a**, **32a**, and **33a**. On the other hand, irradiation in the presence of HCl dramatically increased the yield of *syn*HT dimer **31a** together with suppression of *E/Z*-isomerization. In addition, an increase in the amount of HCl improved the yield of the *syn*HT dimer. Irradiation of **30b** possessing a strongly electron-withdrawing CF_3 group, however, resulted in a decrease in the selectivity even in the presence of HCl. In contrast, the reaction of **30c** having an OMe group yielded *syn*HT dimer **31c** in considerably higher selectivity. These results suggest that the two molecules of **30c** are oriented in a head-to-tail fashion via cation–π interactions prior to the photochemical reaction in acidic conditions, resulting in high regio- and stereoselectivities (Scheme 4.7).

The X-ray crystal structure of **30a**·HCl shows the contribution of pyridinium–π interactions: Two molecules are oriented in a head-to-tail fashion with a double

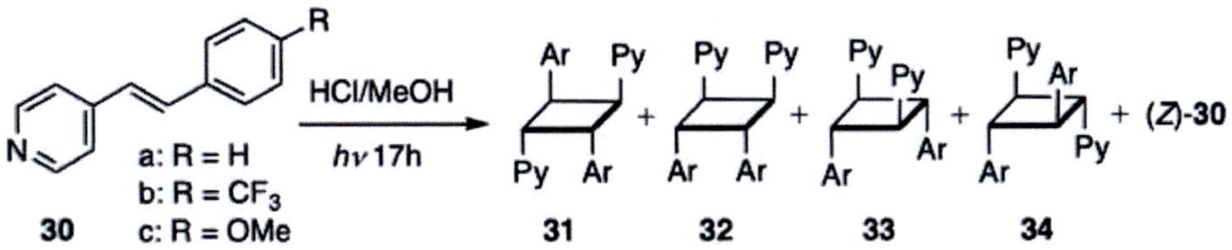

30	HCl (%)	Conv (%)	Products (%)				
			31	32	33	34	(Z)-30
a	0	81	20	14	14	0	52
a	1	91	50	10	14	0	26
a	3	96	71	7	14	0	8
b	3	94	27	24	0	6	43
c	3	94	95	2	0	0	3

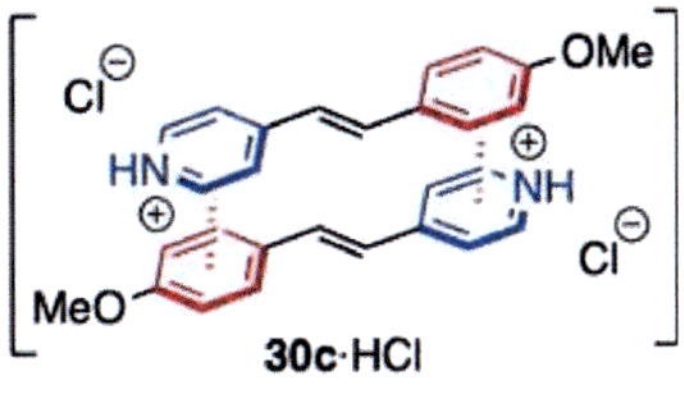

Scheme 4.7 [2+2] Photodimerization of (*E*)-styrylpyridines in the presence of HCl

bond distance of 3.295 Å (Fig. 4.5), supporting the significant role of the cation–π interaction between the pyridinium and benzene rings in [2+2] photodimerization in solution.

Stereoselective [4+4] photodimerization of 1-azanthracene **35** was achieved using a preorientation strategy (Scheme 4.8) [12]. While irradiation of a methanol solution of 1-azanthracene **35** gave a mixture of dimers **36**–**38**, in acidic methanol solutions, *anti*HT dimer **36** was obtained in good yield. By increasing the amount of HCl to 3 equivalents, the *anti*HT dimer **36** was obtained in high yield together with decreases in the *anti*HH and *syn*HT dimers (**37** and **38**).

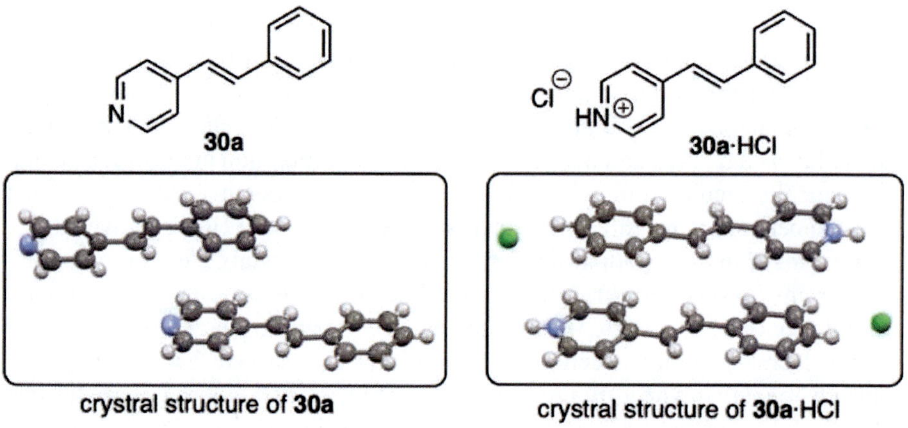

Fig. 4.5 X-ray crystal structures of **30a** and **30a**·HCl. Reproduced with permission from [1]. Copyright 2018 American Chemical Society

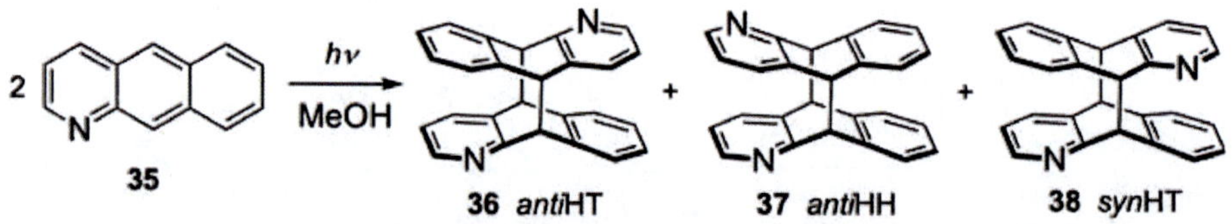

HCl (eq)	Conc (M)	Conv (%)	Products (%) 36	37	38
0	1.0	>99	52	26	22
1	1.0	>99	76	24	0
3	1.0	>99	95	5	0

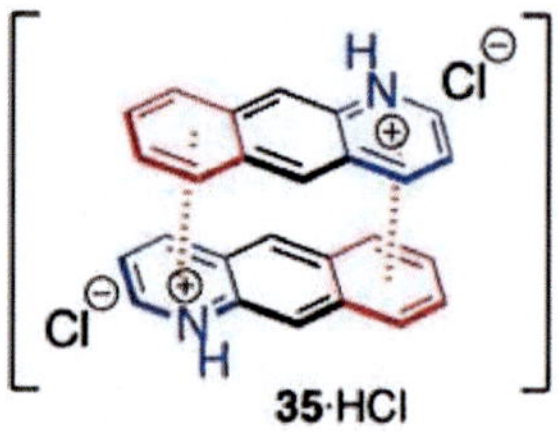

Scheme 4.8 [4+4] Photodimerization reactions of 1-azaanthracene **35** in acidic media. Reprinted with permission from [1]. Copyright 2018 American Chemical Society

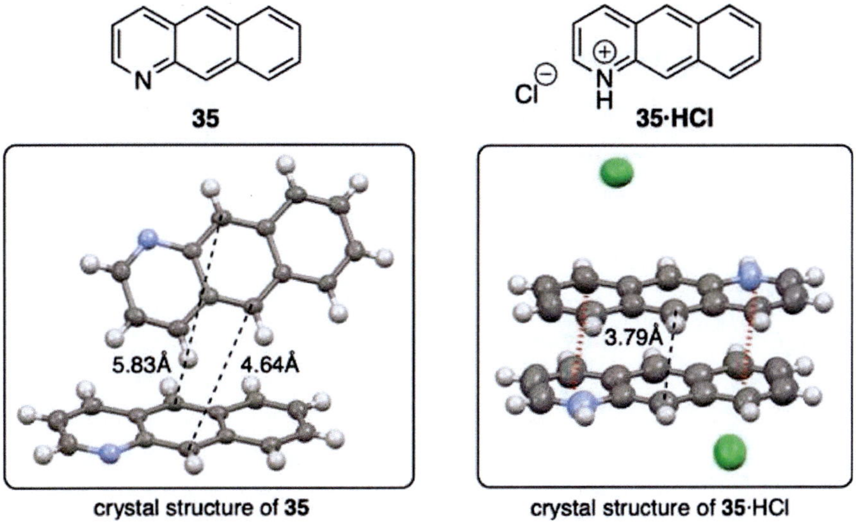

Fig. 4.6 X-ray crystal structures of **35** and **35**·HCl. Reprinted with permission from [1]. Copyright 2018 American Chemical Society

A comparison of the X-ray crystal structures of **35** and **35**·HCl reveals significant differences in their orientation mode (Fig. 4.6). The molecules of **35** are arranged in a non-parallel manner, whereas those of the **35**·HCl salt are oriented parallel to each other in a head-to-tail orientation with 3.79 Å between the two adjacent molecules. These results show the effectiveness of the cation–π-controlled preorientation strategy in photodimerization reactions.

4.2.1.6 Templated-Cyclization

Ring-closing metathesis (RCM) reactions afford a useful method for the synthesis of macrocyclic systems from dialkenes. Collins et al. reported that the RCM reaction of dialkene **39** using Grubbs' catalyst **40** was unsuccessful, but the addition of *N*-methylquinolinium **41** successfully promotes cyclization to give paracyclophane **42** in 45% yield (Scheme 4.9a) [13]. This suggests that *N*-methylquinolinium **41** forms a complex **43**, shielding one side of the benzene ring with a cation–π interaction that orients the two arms of **39** to the same side, making the RCM reaction much easier. Substrate **44**, a two-armed substrate with a terminal alkyne moiety, was similarly cyclized via complex **46** to afford paracyclophane **45** in 42% yield (Scheme 4.9b).

Stoddart et al. prepared a new type of viologen-based tetracationic cyclophane ExBox-4PF$_6$ **51** in 19% yield by reacting **47** and **48** [14]. The yield was greatly improved to 42% when pyrene **49** was used as a template (Scheme 4.10). This may be the result of complex formation between the bromide intermediate **50** and pyrene **49**, which facilitated the intramolecular S$_N$2 reaction of **50** via cation–π interactions. The importance of charge-transfer interactions was noted by the authors.

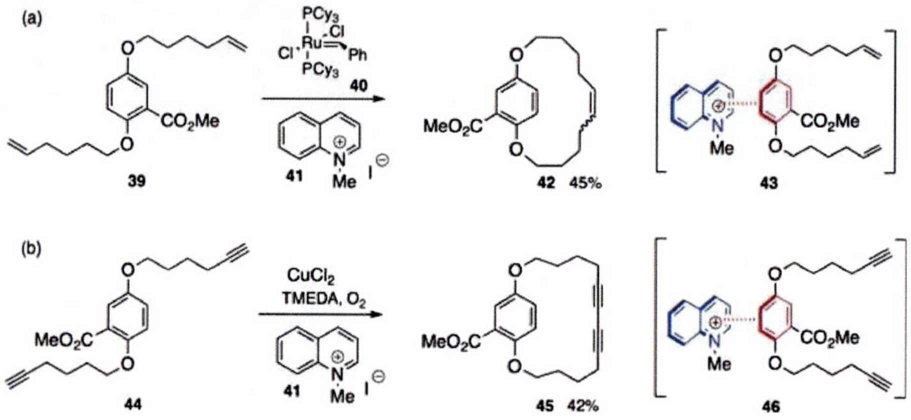

Scheme 4.9 Ring-closing metathesis and Glaser–Hay reactions via templation with *N*-methylquinolinium **41**. Reprinted with permission from [1]. Copyright 2018 American Chemical Society

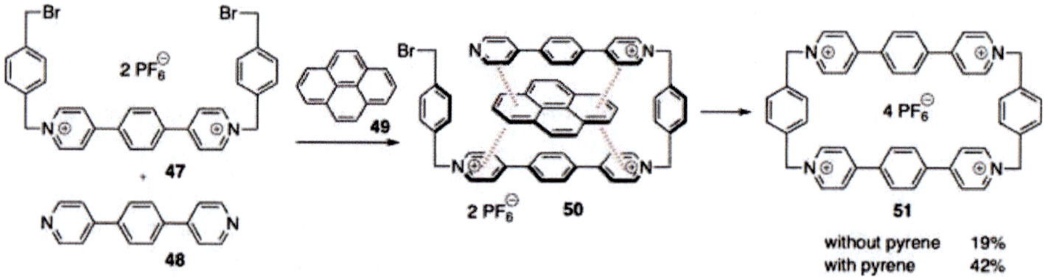

Scheme 4.10 Templated synthesis of cyclophane Exbox· $4PF_6$ **51**. Reproduced with permission from [1]. Copyright 2018 American Chemical Society

4.2.1.7 Anti-Markovnikov Addition

Kumar et al. found that the ionic liquid [bmim]Br was effective in the synthesis of thioethers **54** by a thiol-ene reaction (Scheme 4.11) [15]. The reaction of 4-methylstyrene **52** with thiophenol **53** in [hmim]Br at 40 °C for 30 min gave the anti-Markovnikov product **54** in 80% yield. Studies on the reactions of styrene with various thiols revealed that electron-withdrawing substituents on the thiophenol and benzylthiols required longer reaction times due to their weaker nucleophilicity. In addition, the aliphatic thiol did not undergo hydrothiolation.

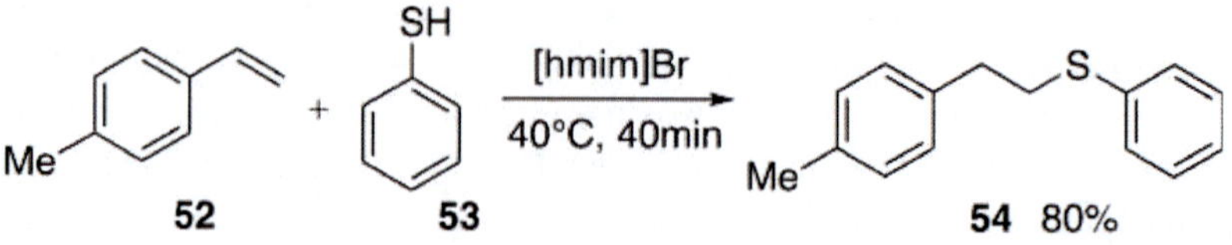

Scheme 4.11 Anti-Markovnikov addition of thiophenol **53** to 4-methylstyrene **52**

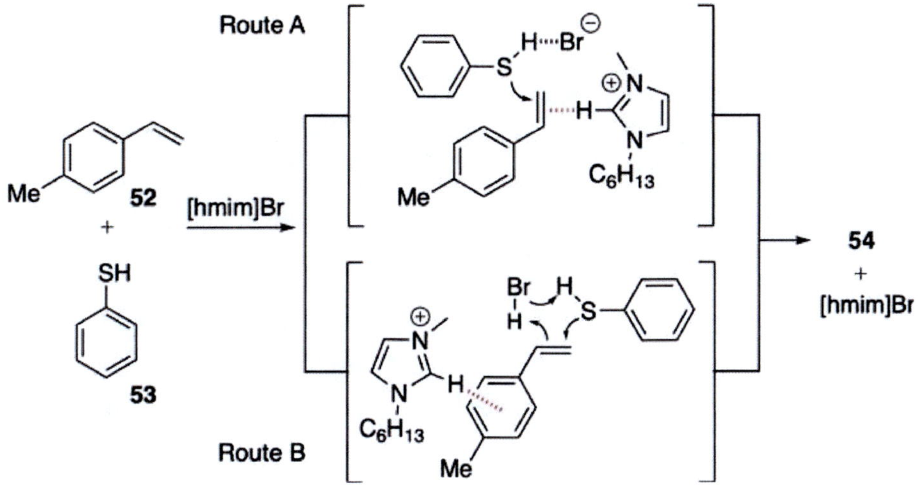

Scheme 4.12 Plausible mechanisms for the anti-Markovnikov addition of **53** to **52**

^{1}H NMR studies predicted the reaction mechanism: When 4-methylstyrene **52** was mixed with [hmim]Br, the C2 proton of the imidazolium ring and the vinyl proton of **52** shifted to lower and higher fields, respectively. These shifts may be due to the activation of **52** by the IL cation through cation–π complex formation and the Br anion activating thiophenol by accepting the proton (route A in Scheme 4.12).

Feng et al. proposed a new anti-Markovnikov one-step addition mechanism for this reaction based on the combination of DFT calculations and experiments [16]. In the new route B, HBr is involved in the reaction through protonation of Br anion as a result of acid–base dissociation of the thiol in the IL or proton exchange of the IL. The free energy of activation for route B is only 20.9 kcal mol^{-1} (Scheme 4.12), which is much lower than that of the Br anion-mediated route A. The reason for the stabilization is considered to be mainly due to the formation of a three-component complex with a six-membered ring.

In the optimized structure of the complex shown in Fig. 4.7, the positively charged imidazolium ring of the hmim is close to the styrene, with the C2-H is brought to face the benzene ring through a T-type cation–π interaction. The benzene ring of the phenylthiol and the imidazolium are also close together, suggesting that these interactions are thought to be important for anti-Markovnikov reactions.

4.2.1.8 Friedel–Crafts-Type Reaction

Narumi et al. identified a Friedel–Crafts-type conjugate addition of indole catalyzed by an azolium salt: In the presence of 10 mol% imidazolium salt **57**, indole **55** attacked chalcone **56** under solvent-free conditions to afford the adduct **58** in 97% yield [17]. The role of the imidazolium salt is thought to be the stabilization of the developing indole anion through cation–π interactions between imidazolium and the

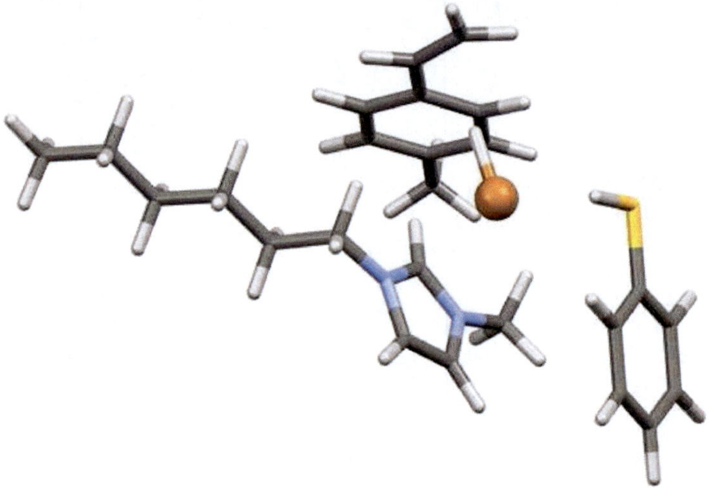

Fig. 4.7 Optimized structure for the complex of imidazolium, 4-methylstyrene, phenylthiol, and HBr

anionic intermediate as shown in Scheme 4.13. High-level ab initio calculations of this complex confirmed the significant contribution of the electrostatic and inductive interactions.

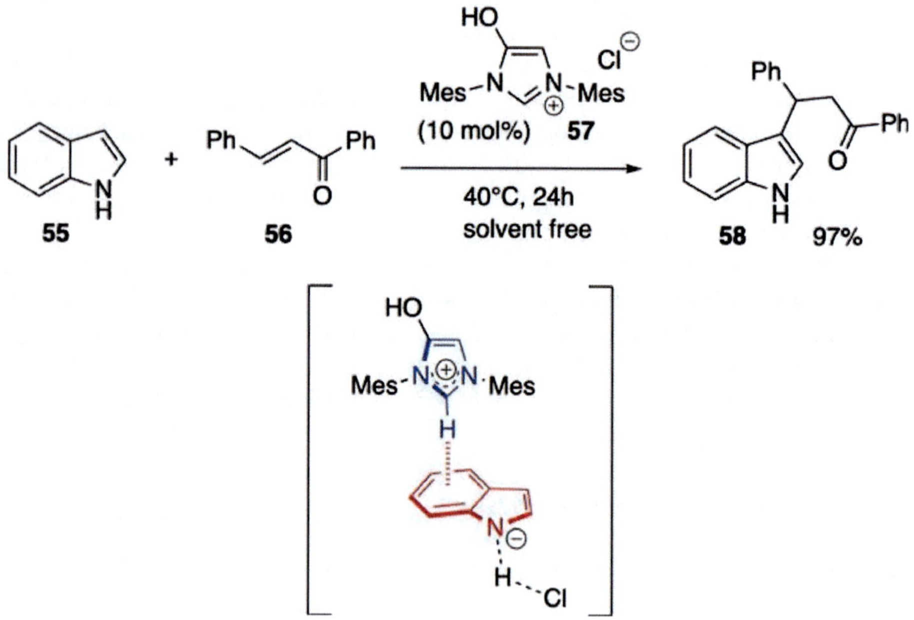

Scheme 4.13 Azolium salt-catalyzed Friedel–Crafts-type conjugate addition of indole. Reprinted with permission from [1]. Copyright 2018 American Chemical Society

4.2.2 *Iminium Ions*

4.2.2.1 Pictet–Spengler Reaction

Comins et al. reported the synthesis of tetrahydroisoquinoline alkaloids using a Pictet–Spengler reaction (Scheme 4.14) [18]. The reaction of chiral carbamates **59** with vinyl ethers gave chiral tetrahydroisoquinolines **60** via an iminium intermediate **62,** and successive reduction of the N-protecting group with $LiAlH_4$ gave (-)-laudanosine **61**. The key to the successful control of the stereocenter is the intramolecular cation–π interaction between the iminium and the benzene ring, where one side of the iminium plane is shielded by the phenyl group of the chiral auxiliary and can be attacked by the benzene ring from the unshielded side, as shown in TS-60. This transition state is similar to that of chiral dihydropyridone synthesis described in Sect 4.2.1.2.

Recently, Scharf and List developed a highly enantioselective catalytic Pictet–Spengler reaction using electron-rich IPPi catalysts that effectively stabilized intermediate iminium ions through cation–π interactions [19]. Using this reaction, naturally occurring tetrahydroisoquinoline alkaloids, such as aporphine, tetrahydroberberine, morphinan, and androcymbine, were synthesized.

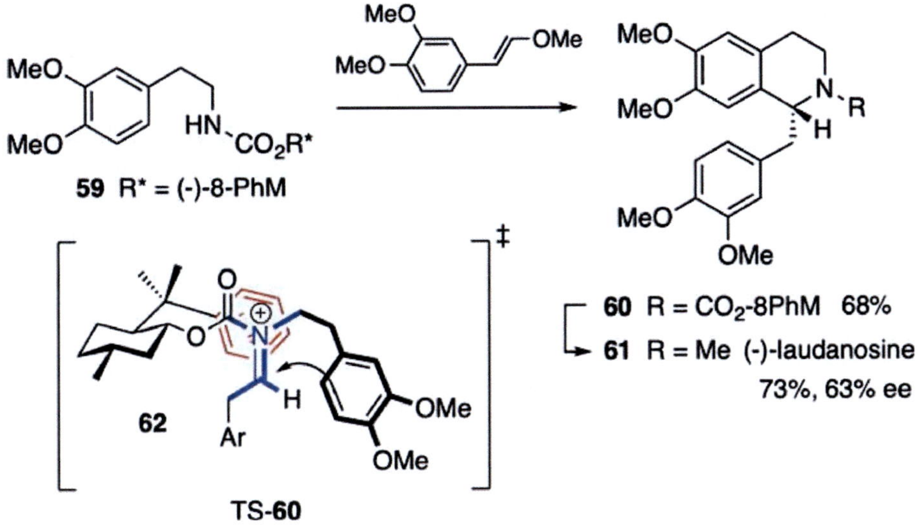

Scheme 4.14 Synthesis of tetrahydroisoquinoline alkaloid using a Pictet–Spengler reaction. Reprinted with permission from [1]. Copyright 2018 American Chemical Society

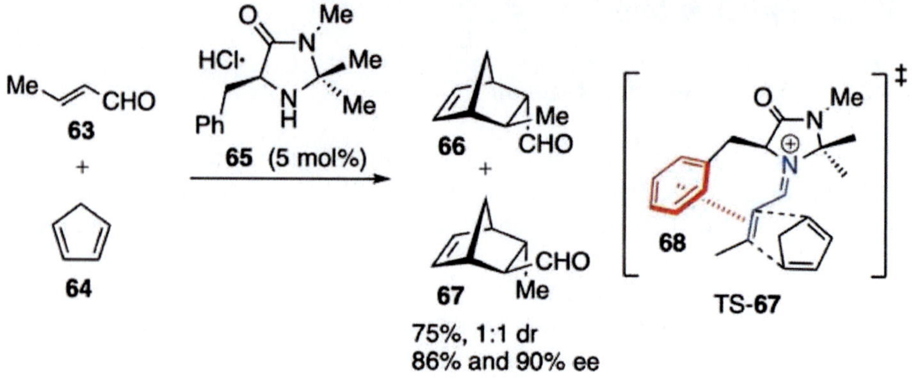

Scheme 4.15 Enantioselective intra- and intermolecular organocatalytic Diels–Alder reactions. Reprinted with permission from [1]. Copyright 2018 American Chemical Society

4.2.2.2 Diels–Alder Reaction

MacMillan's organocatalytic strategy for the activation of unsaturated aldehydes was used in the enantioselective Diels–Alder reaction [20]. In the presence of 5 mol% imidazolidinone catalyst **65**, the cyclization reaction of α, β-unsaturated crotonaldehyde **63** with cyclopentadiene proceeded to give bicyclic aldehydes **66** and **67** in high ee (Scheme 4.15). Based on the optimized structure of the iminium intermediate and energy decomposition analysis by ab initio calculations, the benzyl moiety of the catalyst is thought to effectively shield one side of the iminium plane through a cation–π interaction, exposing the *Si* side to attack by cyclopentadiene **64** via TS-67 [21].

4.2.2.3 Friedel/Michael Reaction

MacMillan et al. also designed the second-generation catalyst **70**, which was more effective than the first-generation catalyst **65** [22]. This catalyst shows improved efficiency for iminium formation as the nitrogen lone pair is almost sterically unaffected. Carbon−carbon bond formation of the intermediate **72** with *N,N*-dimethyl-3-anisidine **69** afforded **71** in high reactivity with 89% ee as the diastereotopic face of the iminium ion **72** is free from steric hindrance, as shown in the transition state TS-71 (Scheme 4.16).

Seebach et al. reported the X-ray crystal structure of the related intermediate iminium salt **73** [23]. The benzyl moiety efficiently shields the (*Z*)-enoyliminium moiety, supporting the transition state described above (Fig. 4.8).

4.2.2.4 Polyene Cyclization

Jacobsen et al. [24]. reported a thiourea-catalyzed asymmetric polyene cyclization reaction to give **76** using hydroxylactam as a cation precursor (Scheme 4.17). A

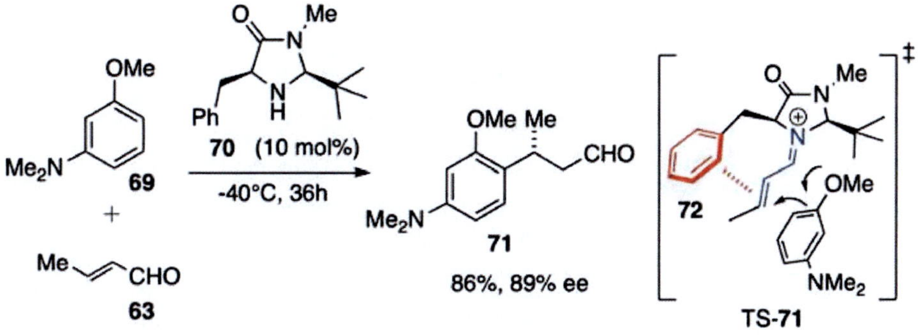

Scheme 4.16 1,4-Addition of an electron-rich aromatic compound to an iminium intermediate. Reprinted with permission from [1]. Copyright 2018 American Chemical Society

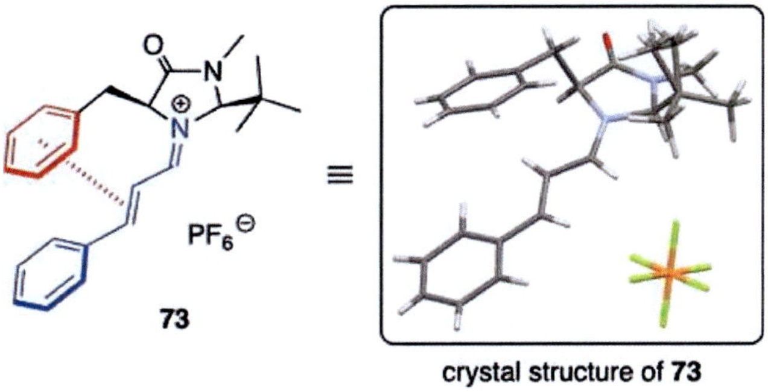

Fig. 4.8 X-ray crystal structure of iminium intermediate **73**. Reprinted with permission from [1]. Copyright 2018 American Chemical Society

survey of catalysts for the cyclization of hydroxylactam **74** revealed that the most effective catalyst was a 4-pyrenyl-substituted thiourea derivative **75**. The fact that the catalyst with a larger π-system resulted in higher enantioselectivity suggests a significant contribution of cation–π interactions, similar to the cation cyclization reactions observed in the biosynthetic process described below (Sect. 4.5). Kinetic studies on these catalytic reactions revealed the correlation between $\Delta\Delta H^{\ddagger}$ and the size of the π-system, indicating the contribution of cation–π interactions.

4.2.3 Oxonium Ions

4.2.3.1 Mukaiyama Aldol Reaction

Jacobsen and coworkers reported Mukaiyama aldol-type reactions catalyzed by chiral squaramide **80** [25]. Squaramide **80** and TBSOTf (50 mol%) catalyzed the reaction of trimethylsilyl enol ether **79** with 4-bromobenzaldehyde dibenzylacetal **78** to afford

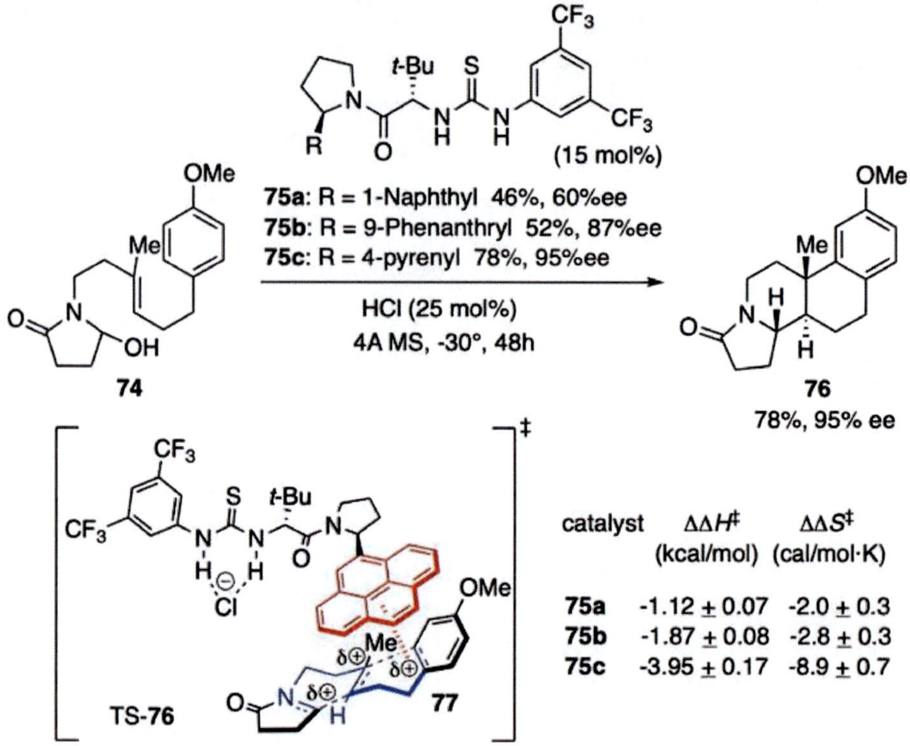

catalyst	$\Delta\Delta H^{\ddagger}$ (kcal/mol)	$\Delta\Delta S^{\ddagger}$ (cal/mol·K)
75a	-1.12 ± 0.07	-2.0 ± 0.3
75b	-1.87 ± 0.08	-2.8 ± 0.3
75c	-3.95 ± 0.17	-8.9 ± 0.7

Scheme 4.17 Thiourea-catalyzed asymmetric polyene cyclization. Reprinted with permission from [1]. Copyright 2018 American Chemical Society

the aldol product **81** (Scheme 4.18). As both reactivity and enantioselectivity are strongly dependent on the size of the aromatic ring on the pyrrolidine ring, the interaction of the aromatic ring of the catalyst with the corresponding cationic intermediate **82** plays a crucial role in this reaction. The role of the aromatic moiety appears to be the stabilization of the oxonium ion of the intermediate and the control of the face selectivity of the nucleophilic reaction via cation–π interactions as described in Sects. 4.2.1 and 4.2.2.

4.2.3.2 cADPR Synthesis

Cyclic ADP-ribose (cADPR) **85** generated from β-NAD **83** by ADP-ribosyl cyclase is known to induce calcium release from intercellular calcium stores. In this enzymatic reaction, Trp77 is proposed to assist the removal of the nicotinamide moiety in the transition state via cation–π interactions, leading to the formation of an oxonium ion intermediate **84** [26]. The adenine base attacks the oxonium ion stabilized by Trp77 from the opposite side to yield **85** (Scheme 4.19).

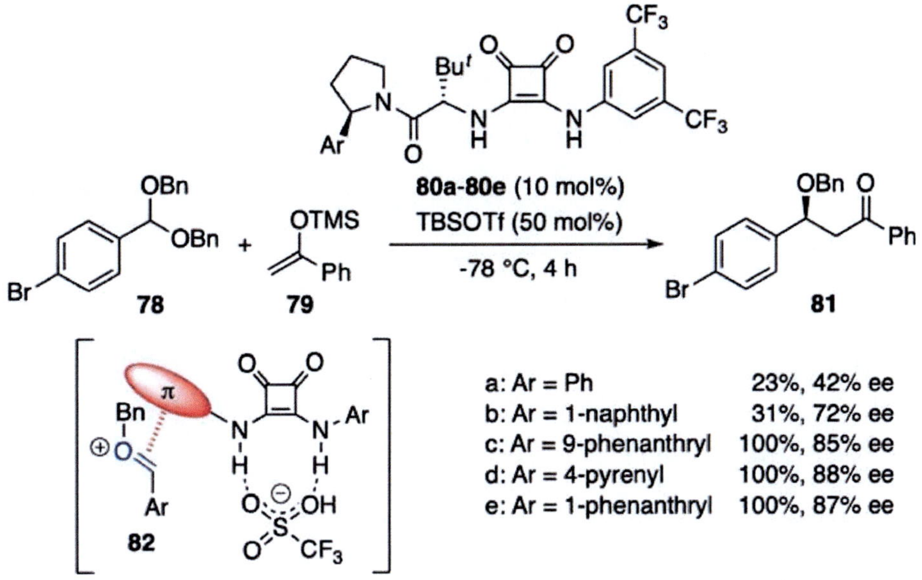

Scheme 4.18 Mukaiyama aldol-type reactions catalyzed by a chiral squaramide in the presence of TBSOTf. Reprinted with permission from [1]. Copyright 2018 American Chemical Society

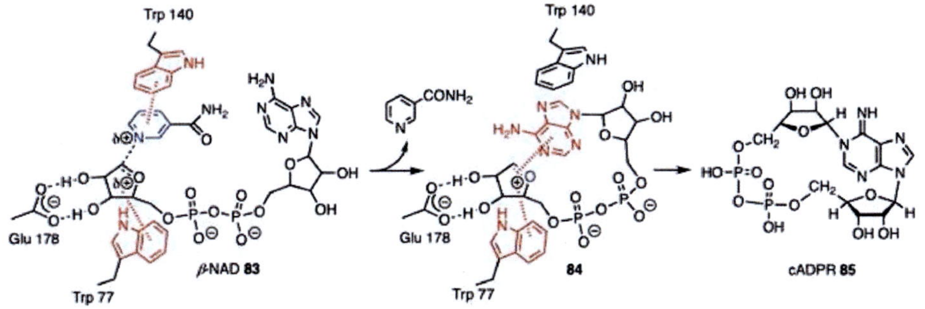

Scheme 4.19 Proposed mechanism for the formation of cADPR by ADP-ribosyl cyclase. Reproduced with permission from [1]. Copyright 2018 American Chemical Society

The cyclization of NAD$^+$ **83** to cADPR **85** was achieved by chemical reactions [27]. When β-NAD$^+$ was heated in DMSO in the presence of NaBr, cADPR was obtained in 28% yield (Scheme 4.20). α-NAD$^+$ **86** was also cyclized to give the same compound in 17% yield under similar reaction conditions, suggesting that the reaction proceeds via oxocarbenium ion **84** in a similar manner to the enzymatic reaction shown in Scheme 4.20. The adenine base seems to stabilize the oxocarbenium ion as well as the transition state for the formation of the ion through a cation–π interaction, and the adenine base then attacks the oxocarbenium ion to produce cADPR (Scheme 4.20).

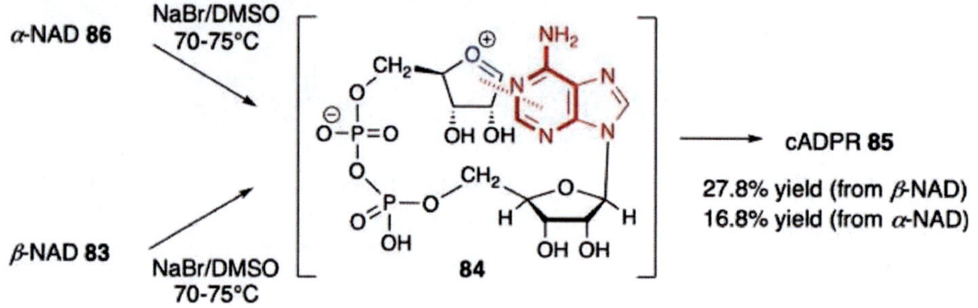

Scheme 4.20 Biomimetic cyclization of β-NAD^+ and α-NAD^+. Reprinted with permission from [1]. Copyright 2018 American Chemical Society

4.2.4 *Carbocations*

4.2.4.1 Friedel–Crafts Reaction

Denis et al. prepared the 3, 3-diarylazetidines **89** from *N*-Cbz-azetidinol **87** and *o*-methylphenol **88** by a calcium-catalyzed Friedel–Crafts reaction [28]. Studies on the reactions of various N-substituted compounds revealed that substrates with *N*-Boc, *N*-CO_2Me, and *N*-COMe groups gave no adduct, but the substrate possessing a *N*-Cbz group effectively afforded product **89**. The remarkable role of the Cbz group may be due to the stabilizing effect of the intermediate carbocation **90** through a cation–π interaction, as shown in Scheme 4.21.

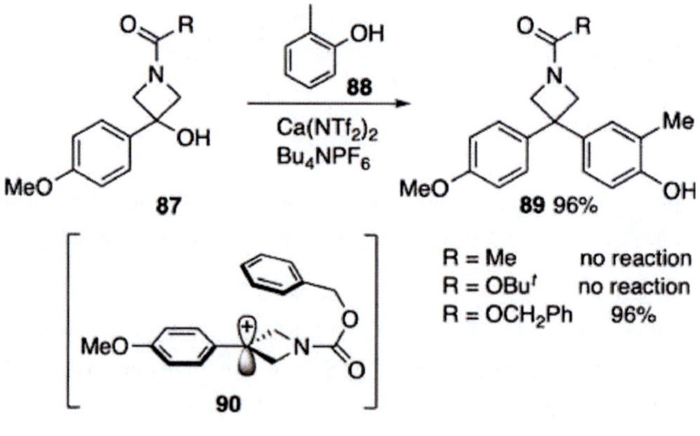

Scheme 4.21 Effect of the Cbz group on the Friedel–Crafts reaction of azetidinol and *m*-cresol in the presence of $Ca(NTf_2)_2$ and Bu_4NPF_4

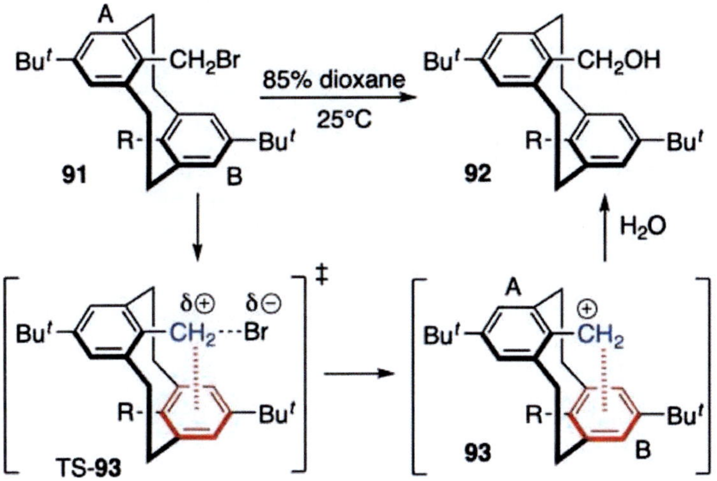

R	k x 10^7 (sec^{-1})
H	62
Me	550
OMe	2920
F	272
CN	199

Scheme 4.22 Substituent dependence of the hydrolysis of 8-bromomethyl [2.2]metacyclophanes. Reprinted with permission from [1]. Copyright 2018 American Chemical Society

4.2.4.2 Hydrolysis of Alkyl Halides

Hydrolysis of 8-bromomethyl[2.2.2]metacyclophane **91** to the corresponding 8-hydroxymethyl derivative **92** was carried out in 85% dioxane solution at rt (Scheme 4.22). The rate of hydrolysis of the bromomethyl moiety on the aromatic ring **A** was strongly dependent on the substituent attached to the other aromatic ring **B** of the cyclophane [29]. The presence of electron-withdrawing substituents at the benzene ring B decreased the rate of hydrolysis, whereas the presence of electron-donating substituents increased it. Based on these observations, the authors speculate that the aromatic ring B stabilizes both the transition state TS-93 for carbocation formation and the intermediate carbocation **93** by an intramolecular cation–π interaction.

4.2.5 Quaternary Ammonium Ions

4.2.5.1 Norrish–Yang Reaction

The Norrish–Yang reaction of α-alkoxybenzophenone **94** generally produced *cis*-dihydrobenzofuran derivative **96**. Yamada et al. found that the addition of tetrabutylammonium (TBA^+) changed the stereoselectivity, giving the *trans* isomer **95** as the major product [30]. The reaction of substrate **94a** with a methoxy group showed higher *trans* selectivity than that of **94b** with a CF_3 group (Scheme 4.23). Ab initio calculations for the complex between the substrate and TBA^+ showed that TBA^+ and both the carbonyl group and the benzene ring of the benzyl moiety are

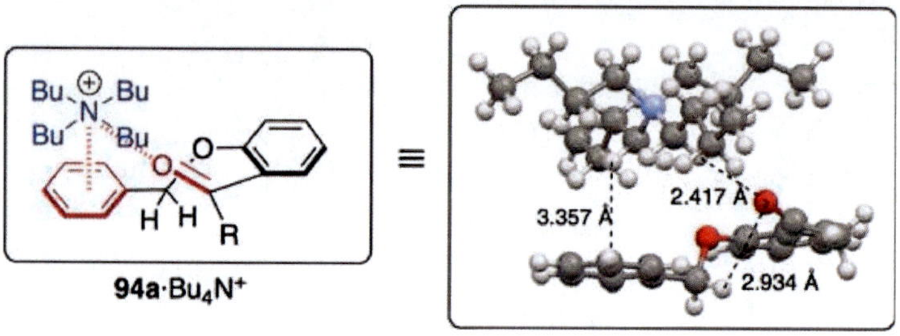

Fig. 4.9 Optimized structure of a complex of substrate **94a** and TBA$^+$. Reprinted with permission from [1]. Copyright 2018 American Chemical Society

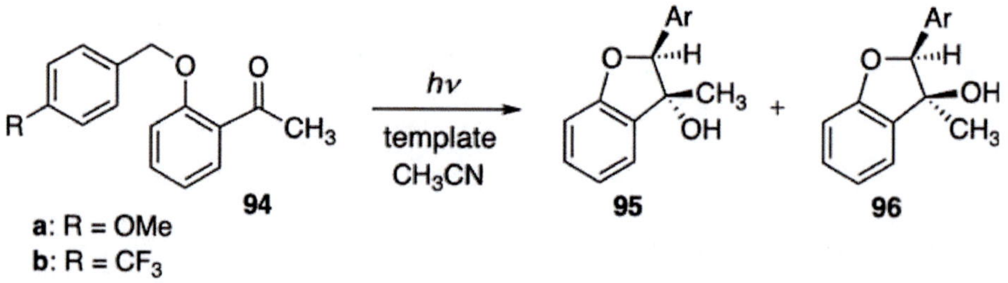

R	template (5eq)	time (h)	conv (%)	**95 : 96**
OMe	–	24	70	11:89
OMe	Bu_4NF	24	99	94:6
CF_3	–	24	33	23:77
CF_3	Bu_4NF	24	6	59:41

Scheme 4.23 Tetrabutylammonium ion-assisted Norrish–Yang cyclization of acetophenone derivatives. Reprinted with permission from [1]. Copyright 2018 American Chemical Society

close together, suggesting the contribution of cation–π interactions (Fig. 4.9). This complex formation is expected to lead to the *trans* selectivity.

4.2.5.2 Michael Addition

Paton et al. [31] reported an intramolecular stereoselective cyclization catalyzed by chiral tetraalkylammonium. Treatment of malonic acid derivatives **97** with a base in the presence of chiral tetraalkylammonium ions **98** resulted in 5-endo-trig cyclization, yielding substituted indenes **99** with high enantio- and diastereoselectivities (Scheme 4.24). The proposed transition state TS-99 exhibits interactions between the

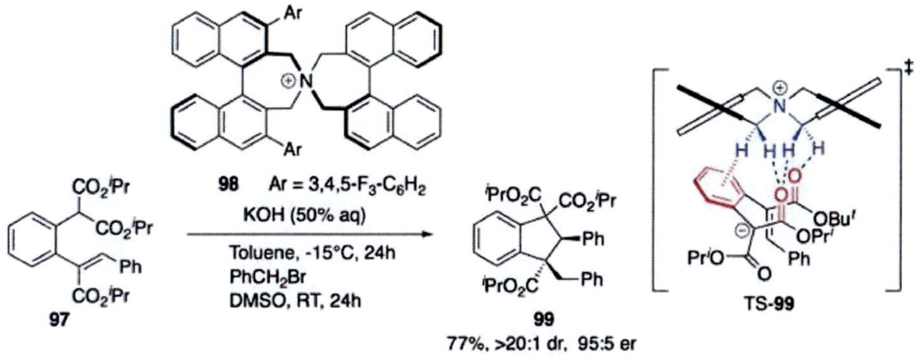

Scheme 4.24 Chiral tetraalkylammonium-assisted intramolecular Michael reaction. Reprinted with permission from [1]. Copyright 2018 American Chemical Society

ammonium and the aromatic ring and two carbonyl groups, leading to high stereoselectivity. Similar interactions have been reported in the TBA^+-assisted Norrish–Yang reaction described in the previous section.

4.2.5.3 Alkyne Cyclization

Yasuhara and Sakamoto found that the cyclization of 2-ethynylbenzylalcohol **100** gave the isobenzofuran derivative **101** in good yield (Scheme 4.25a), in which a cation–π interaction between the triple bond and TBA^+ enhanced the cyclization of the alkoxy moiety via TS-101 [32]. This method was applied to the cyclization of alkynylanilines and alkynylphenols for the synthesis of carbazole derivatives and benzo[b]furan natural products such as vinafuran, cumestane, and coumestrol.

Fujita et al. reported the TBAF-catalyzed carboxylative cyclization of propargylic amine **102** by reaction with CO_2 [33]. The stirring of a *t*-BuOH solution of propargylic amine **102** for 12 h at 100 °C in the presence of 0.5 mol % of TBAF in an autoclave under a CO_2 atmosphere of 1.0 MPa afforded the corresponding 2-oxazolidinone **104** in 85% yield through an intermediate **103** (Scheme 4.25b).

Dihydroisobenzofuran derivative **106** is prepared by a similar cyclization. The TBAF-catalyzed selective O-nucleophilic cyclization of enaminone **105** was achieved via isomerization of the enaminone moiety by deprotonation with a fluoride ion and activation of the triple bond through a cation–π interaction with TBA^+ [34] (Scheme 4.25c).

Maity and Lepore reported a TBAX-catalyzed alkyne cyclization of β-alkynylhydrazines [35]. This reaction was applied to the kinetic resolution of propargylic esters **107** using chiral ammonium fluoride **108**. Treatment of substrate **107** with 10 mol% chiral catalyst afforded dehydroazaproline **109** in 48% yield, allowing recovery of the chiral substrate **107** in 93% ee (Scheme 4.26).

Raman spectroscopic studies revealed the role of tetraalkylammonium in accelerating the rate of the cyclization reaction [36]. As the amount of TBA^+ is increased,

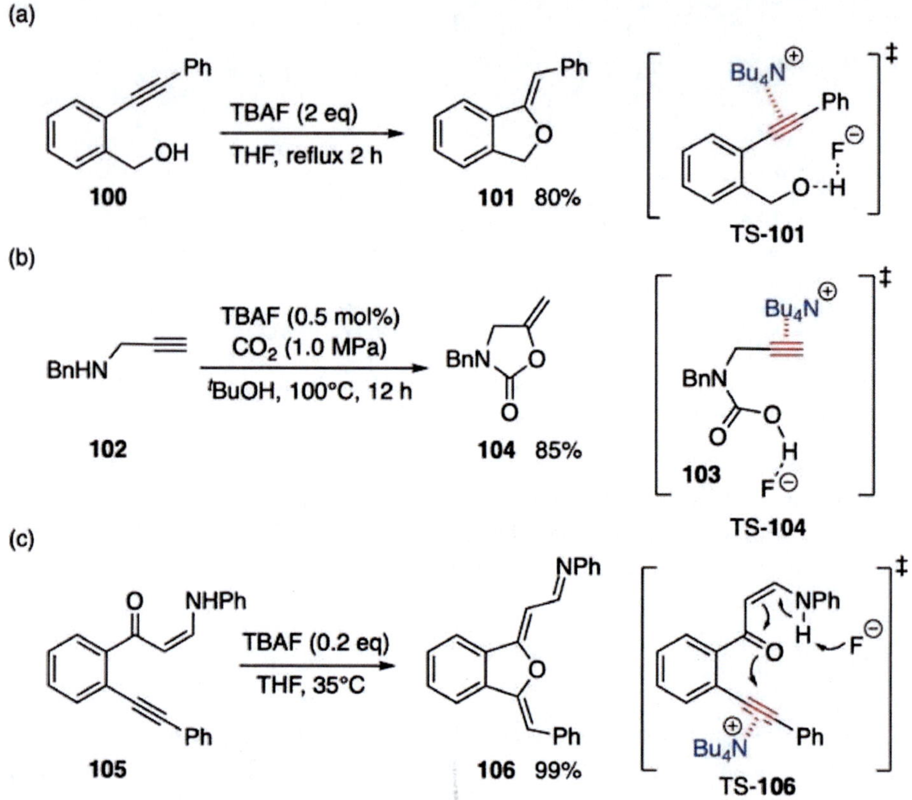

Scheme 4.25 TBAF-mediated cyclization of **a** 2-ethynylbenzyl alcohol 100, **b** propargylic amine 102 with CO_2, and **c** enaminone 106. Reprinted with permission from [1]. Copyright 2018 American Chemical Society

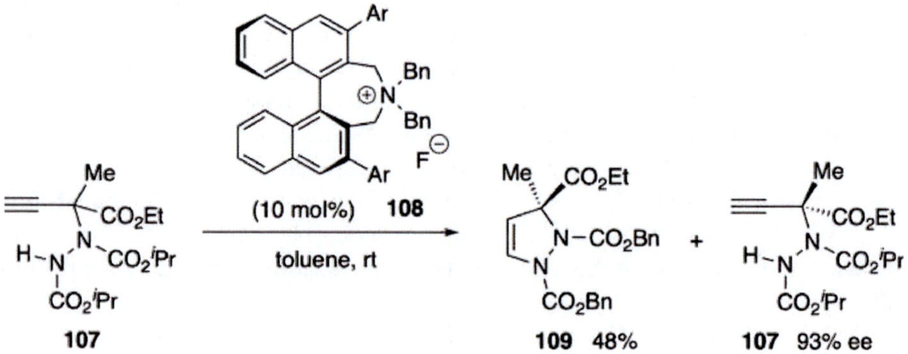

Scheme 4.26 Chiral tetraalkylammonium-mediated cyclization reaction of β-alkynyl hydrazine. Reprinted with permission from [1]. Copyright 2018 American Chemical Society

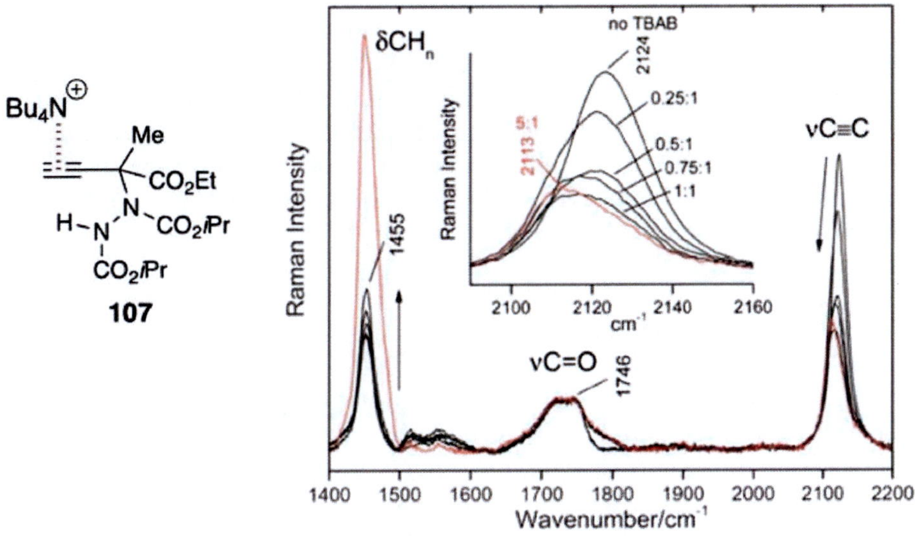

Fig. 4.10 Raman spectra of **107** in the presence of tetrabutylammonium. Reproduced from Ref. [35] with permission from the Royal Society of Chemistry

the absorption band of the C≡C bond shifts to a lower wavenumber (Fig. 4.10). This indicates that TBA$^+$ makes the triple bond more electrophilic by the cation–π interaction between the alkyne and TBA$^+$.

4.2.5.4 Reduction

Yang et al. reported the diastereoselective reduction of ketone **110** with tetraalkylammonium borohydride in the course of the total synthesis of Maoecrystal V [37]. Reduction of ketone **110** with sodium borohydride ($NaBH_4$) in methanol produced a 2:3 mixture of alcohols **111** and **112**, whereas treatment of **110** with Bu_4NBH_4 and

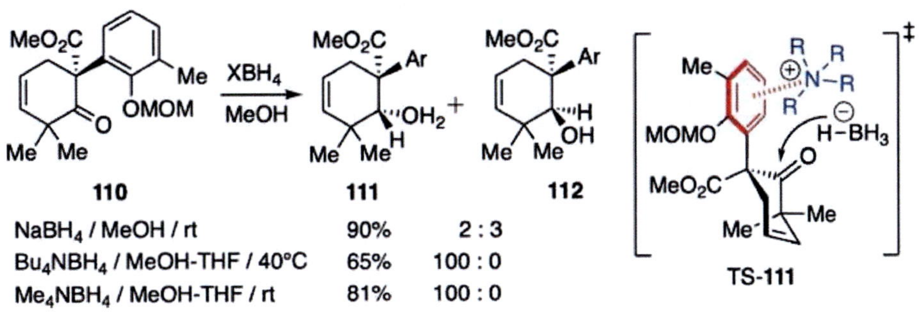

Scheme 4.27 Diastereoselective reduction of ketone with tetraalkylammonium borohydride. Reprinted with permission from [1]. Copyright 2018 American Chemical Society

Me_4NBH_4 in methanol/THF at rt gave the desired alcohol **111** as a single isomer in 65% and 81% yields, respectively (Scheme 4.27).

The very high diastereoselectivities when using Bu_4NBH_4 and Me_4NBH_4 are attributed to the cation–π interaction between tetraalkylammonium and the phenyl ring of the substrate **110**, which leads to an attack by BH_4^- from the same side of the ammonium ion. In other words, the R_4N^+ ion assists the delivery of hydride to the *Si* face of the ketone via TS-111.

4.2.5.5 Dynamic Combinatorial Synthesis

Sanders et al. developed a template-assisted synthetic strategy, involving reversible imine bond formation, to provide thermodynamically stable products [38]. When building block **113** was treated with 1 equiv of TFA in $CHCl_3$, dimer **114** was obtained in 88% yield (Scheme 4.28). On the other hand, in the presence of template **116**, dimer **114** and trimer **115** were produced in 41% and 56% yields, respectively. Interestingly, acetylcholine chloride **117** promoted the formation of trimer **115** under the same reaction conditions (86% yield). It is speculated that the C–H–O hydrogen bond between the CH_2 group of the ammonium and the C=O oxygen atom of the amide is the driving force behind the templating. In addition to these interactions, cation–π interactions between the ammonium and the aromatic ring are also thought

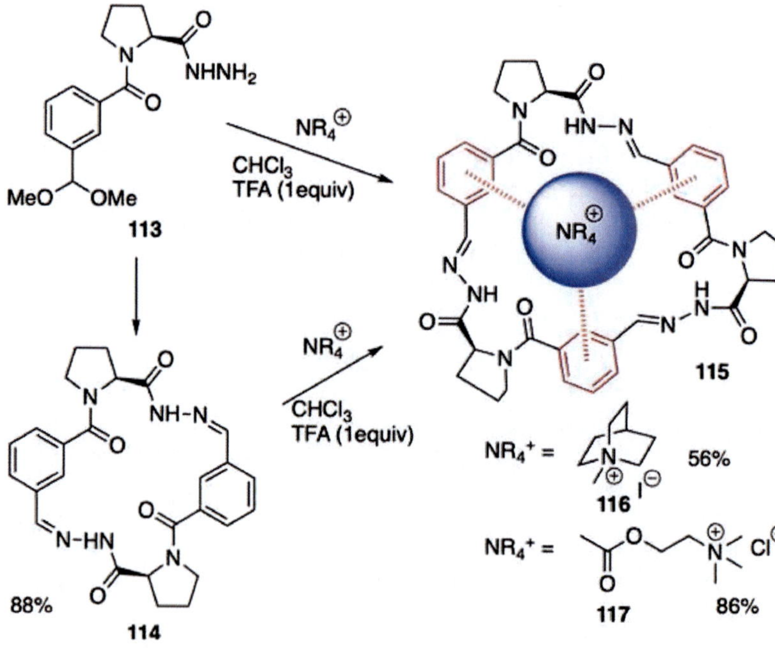

Scheme 4.28 Templated synthesis of cyclic trimers by reversible imine bond formation. Reprinted with permission from [1]. Copyright 2018 American Chemical Society

to play an important role, similar to the case of the Norrish–Yang cyclization reaction (Scheme 4.23) described above.

4.3 Alkali Metal Ions

As described in Chap. 2, experimental evaluation of the interaction energies between alkali metal ions and benzene rings shows that the interaction energy decreases with increasing ionic radius. Three types of interactions between alkali metal ions and a benzene ring are shown in Fig. 4.11; (a) and (b) represent the interaction of benzene with counter cations of enolates and alkoxides, respectively, and (c) represents the interaction between benzene and alkyl metal compounds.

4.3.1 Lithium Ion

Posner et al. observed significant regioselectivity in the synthesis of enol silyl ethers of the 3-methyl-3-arylcyclopentanone derivative **118** using LDA as a base [39]. When the substituent was a *p*-methoxyphenyl group, a 1:7.4 mixture of **119b** and **120b** was formed. In contrast, a ratio of 1:0.9 was observed for the reaction of **118c** with a *p*-nitrophenyl group (Scheme 4.29a). Moreover, much lower selectivity was observed for the reaction of **118d** without a benzene ring. The regioselectivity was dependent on the solvent properties, with increased solvent coordination ability resulting in decreased stereoselectivity (in the order DME < THF < Et_2O < pentane). These observations can be explained by the formation of the $Li^+\cdots\pi$ complex **121** in the transition state TS-120.

Furthermore, diastereoselectivity was also observed in the methylation of enol silyl ethers with methyl iodide. This may be due to the interaction between the phenyl group at the C3 and Li^+, which blocked one side of the enolate face, allowing the reaction with methyl iodide from the opposite side and resulting in the preferential formation of *trans*-product **122**. The addition of CuMe inhibits this interaction to give *cis*-product **123** (Scheme 4.29b).

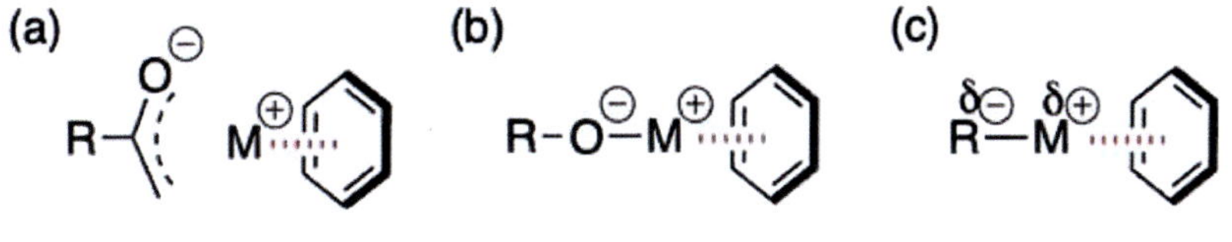

Fig. 4.11 Three types of interactions between alkali metal ions and a benzene ring

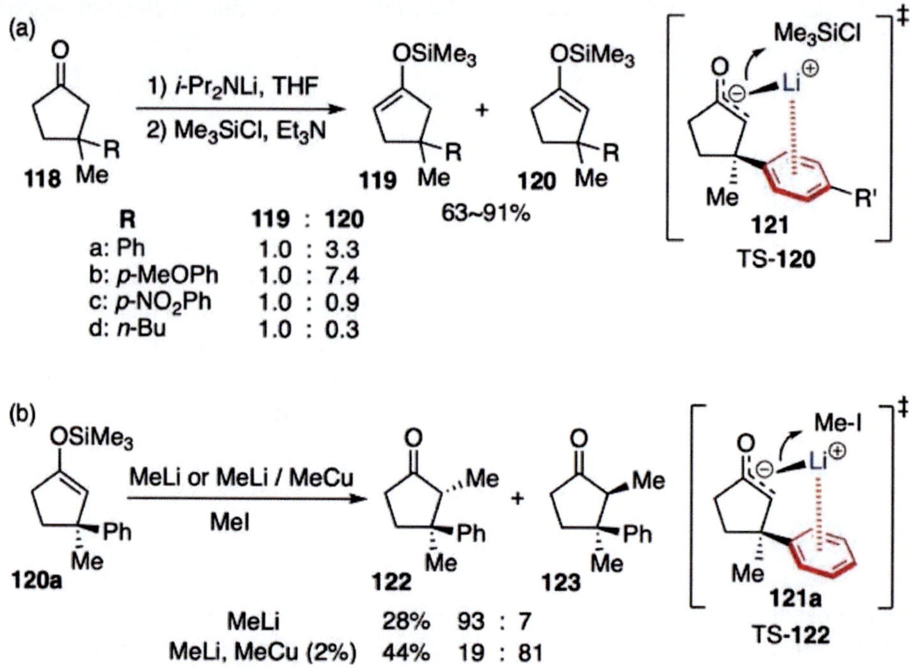

Scheme 4.29 **a** Regioselective synthesis of enol silyl ether and **b** stereoselective methylation of cyclopentanone derivatives through $Li^{+}\cdots\pi$ interactions. Reprinted with permission from [1]. Copyright 2018 American Chemical Society

4.3.2 Sodium Ion

The α-alkylation of amidoenolates generated from **124a** having *trans* 4-phenyl-2-trifluoromethyloxazolidine as a chiral auxiliary produced **125a** in very high diastereoselectivity (Scheme 4.30) [40]. DFT calculations for the transition state of the reaction of the enolate with ethyl iodide suggest that the cation–π interaction between Na^{+} and the phenyl group leads to the attack by the ethyl group from the *Re* face (Fig. 4.12). This is consistent with the lower selectivity in the reaction of **124b**.

Scheme 4.30 Diastereoselective α-benzylation of a chiral amide. Reprinted with permission from [1]. Copyright 2018 American Chemical Society

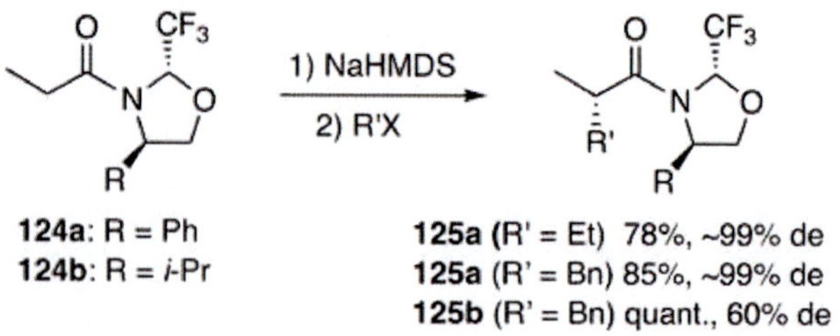

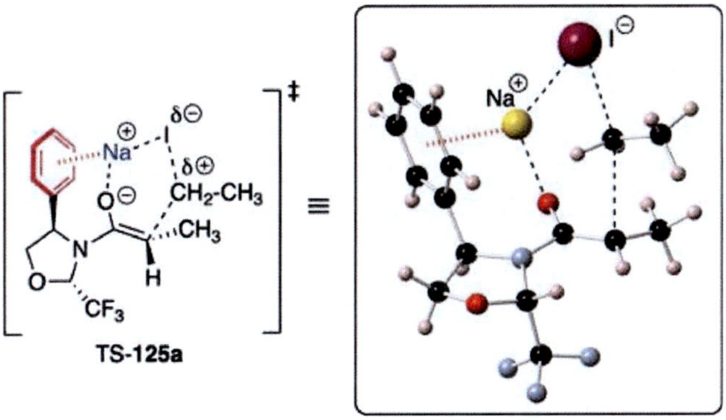

Fig. 4.12 Calculated transition structure for the ethylation of the sodium enolate. Reprinted with permission from [1]. Copyright 2018 American Chemical Society

4.3.3 Potassium Ion

Chiba, Hirao, and coworkers described the potassium ion-mediated hydroamination of alkenyl oxime for the synthesis of cyclic nitrones [41]. In the absence of K_3PO_4, the cyclization of **126** does not proceed, whereas in the presence of K_3PO_4, the reaction smoothly proceeds to yield cyclic nitrones **127** in high yield. DFT calculations for this transition state show that the energy barrier in the presence of K^+ is much lower than that in the absence of K^+, suggesting that the cation–π interaction between K^+ and the benzene ring brings the two reaction centers closer together and contributes to stabilizing the transition state (Scheme 4.31).

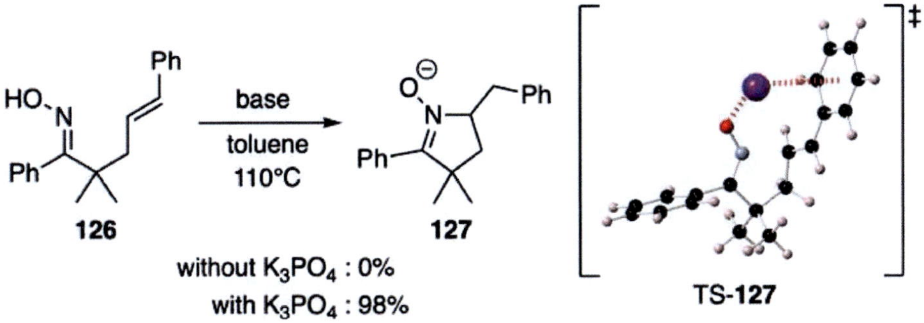

Scheme 4.31 Potassium ion-mediated hydroamination of alkenyl oxime. Reprinted with permission from [1]. Copyright 2018 American Chemical Society

4.4 Supramolecular Systems

As described in Chap. 3, cation–π interactions play an important role in biological systems, and their importance was also recognized early with regard to host–guest systems. This section introduces various examples of organic reactions in artificial supramolecular systems.

4.4.1 *Cyclophanes*

Dougherty et al. reported that the cyclophane host **129** acts as a biomimetic catalyst for the Menshutkin reaction of quinoline **128** in aqueous media to give *N*-methylquinolinium **130** [42]. In comparison with the absence of the host, the addition of 20 mol% of the host significantly increased the reaction rate. This may be due to stabilization of the positive charge developed in the transition state TS-130 by the aromatic ring of host **129** (Scheme 4.32).

Richeter and Rebek developed the resorcinarene-based supramolecular catalyst **131** as shown in Scheme 4.33 [43]. The hydrolysis of carbonate **133** was promoted by catalyst **131**. The reaction rate was accelerated with increases in the amount of the catalyst, confirming the role of cyclophane as a catalyst. When zinc catalyst **132**,

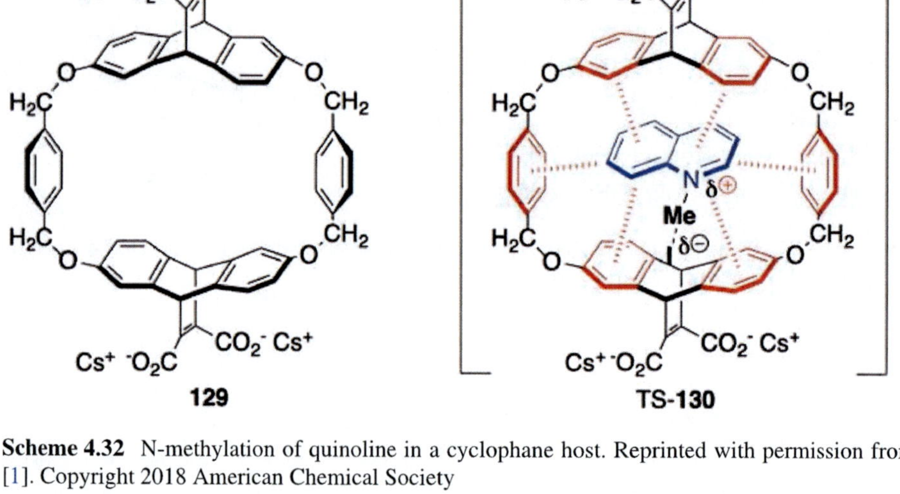

Scheme 4.32 N-methylation of quinoline in a cyclophane host. Reprinted with permission from [1]. Copyright 2018 American Chemical Society

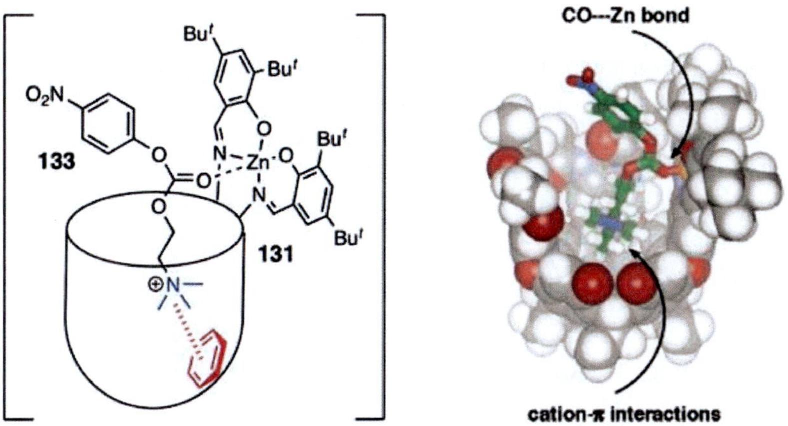

Fig. 4.13 Energy-minimized structure (CaChe 4.9) of the complex between catalyst **131** and substrate **133** through a cation–π interaction. Reprinted with permission from [43]. Copyright 2004 American Chemical Society

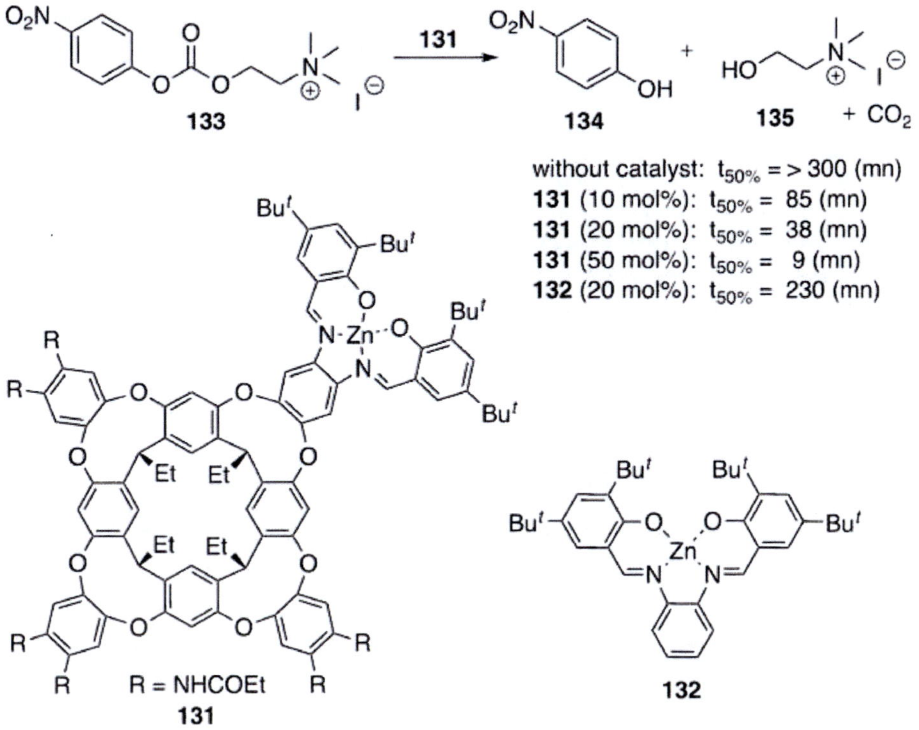

Scheme 4.33 Catalytic hydrolysis of *p*-nitrophenylcholine carbonate with cavitand **131** possessing a chelated zinc ion. Reprinted with permission from [1]. Copyright 2018 American Chemical Society

without the cyclophane moiety, was used, the reaction rate was slower than when **131** was used. Catalyst **131** is thought to recognize *p*-nitrophenylcholine carbonate **133** and to fix it in manner suitable for hydrolysis through coordination of the carbonyl to the zinc ion and the cation–π interaction between the ammonium ion and the aromatic ring of the cavity as shown in Fig. 4.13.

4.4.2 Supramolecular Cages

4.4.2.1 Cationic Cages

Fujita et al. developed a coordination nanocage **136** consisting of four triangular tris(4-pyridyl)triazines, four tris(3-pyridyl)triazines, and six palladium ions (Fig. 4.14) [44]. This positively charged cage provides the space required for the [2+2] and [4+2] cycloaddition and catalytic oxidation reactions of various olefins.

Stereoselective [2+2] photodimerization of acenaphthylene **137** was achieved to yield *syn* dimer **138** in the cage (Scheme 4.34). Crystal-structure analysis of the cage involving **137** revealed that the acenaphthylene molecules overlap in a face-to-face manner on the triazine ligand of the cage at a distance of 3.3–3.5 Å (Fig. 4.15), indicating the contribution of the interaction between the ligand with the positive charge and the electron-rich acenaphthylene. On the other hand, the distance between the double bonds is 8.3–9.0 Å, which is much longer than Schmidt's requirement of 4.2 Å. Nevertheless, within this restricted space, the [2+2] photocycloaddition reaction occurs efficiently to yield the *syn* HH dimer **138**, demonstrating the remarkable catalytic properties of this nanocage.

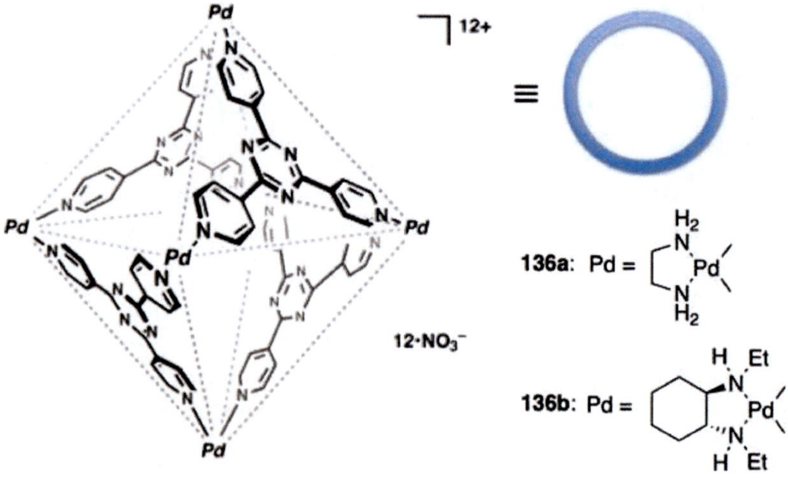

Fig. 4.14 Coordination nanocages. Reproduced with permission from [44]. Copyright 2002 American Chemical Society

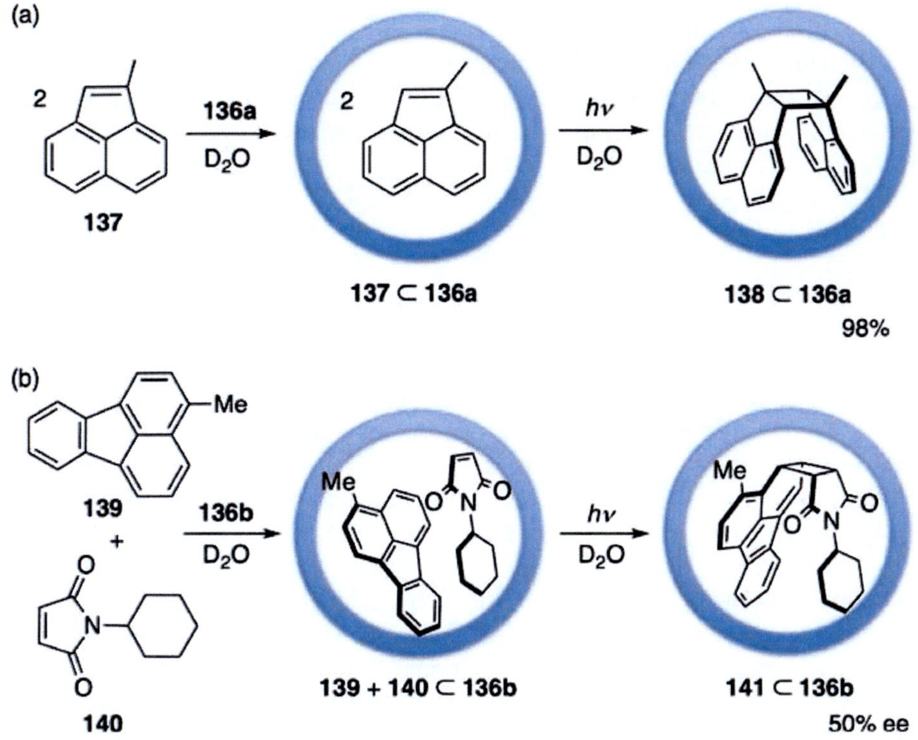

Scheme 4.34 **a** [2+2] Photodimerization of acenaphthylene and **b** enantioselective cross-photocycloaddition of fluoranthene in nanocages. Reprinted with permission from [1]. Copyright 2018 American Chemical Society

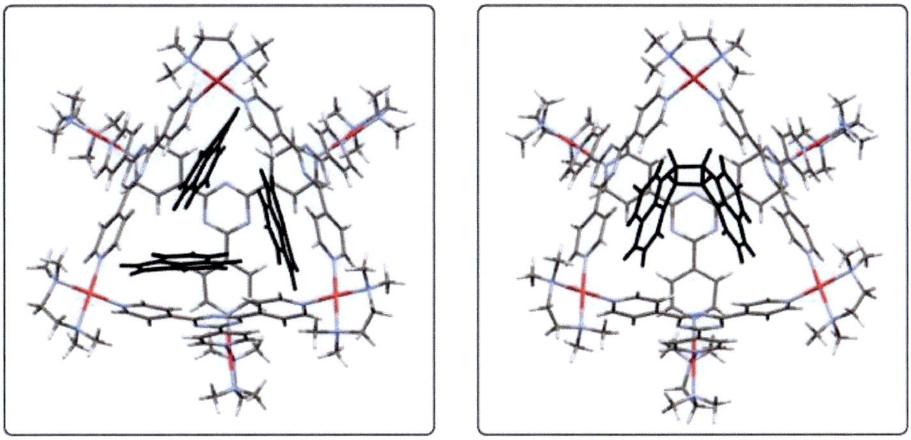

Fig. 4.15 X-ray crystal structures of **137** ⊂ **136a** (left) and **138** ⊂ **136a** (right). Reprinted with permission from [1]. Copyright 2018 American Chemical Society

A chiral cross dimer was obtained by using chiral molecular flask **136b** [45]. When chiral cage **136b** with a chiral diamine ligand on Pd(II) was used, the chiral *syn* cross dimer **141** was produced from **139** and **140** at 50% ee. This suggests the preferential orientation in one of the two enantiomeric orientations of the substrate in the chiral cavity.

4.4.2.2 Anionic Cages

Raymond and coworkers developed the tetrahedral aromatic supramolecular cage $K_{12}Ga_4L_6$ (L=*N*,*N*-bis(2,3-dihydroxybenzoyl)-1,5-diaminonaphthalene) **142a-142c**, which is generated via self-assembly of the ligands and gallium ions (Fig. 4.16) [46]. The Aza–Cope rearrangement of enammonium ion **143** took place within the tetrahedral aromatic supramolecular cage Ga_4L_6. When substrates **143a** (R=H), **143b** (R=Et), and **143c** (R=*i*-Pr) were encapsulated by supramolecular assemblies **142a**, the rates of rearrangement were increased 5-, 141-, and 854-fold, respectively, compared to the reaction rate without the catalyst (Scheme 4.35). The NOESY spectra of [**143b** $\subset$ Ga_4L_6] showed that the two methyl groups of the isobutenyl and ethyl moieties were located in close proximity, showing that the conformation of the substrate was locked into a reactive conformation in the cage. The catalytic cycle of this rearrangement in the cage was proposed as shown in Scheme 4.35. When chiral cage **142b** was used for the rearrangement of **143c, 145c** was obtained in 78% ee [47].

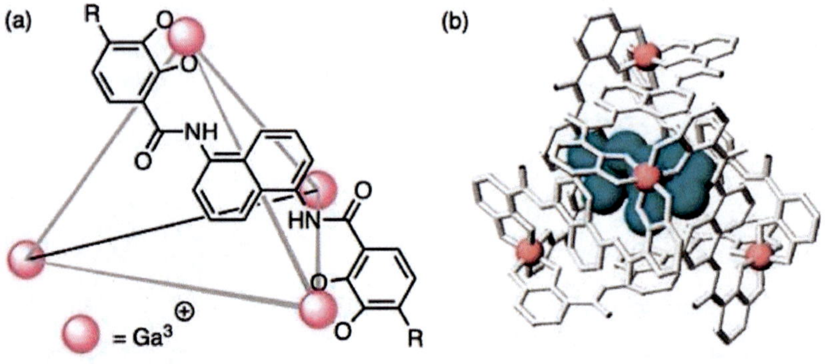

racemic **142a**: R = H
ΔΔΔΔ-**142b**: R = H

Fig. 4.16 Tetrahedral supramolecular hosts $K_{12}Ga_4L_6$ **a** and the complex of **142** $\supset$ **143 b**. Reprinted with permission from [46]. Copyright 2006 American Chemical Society

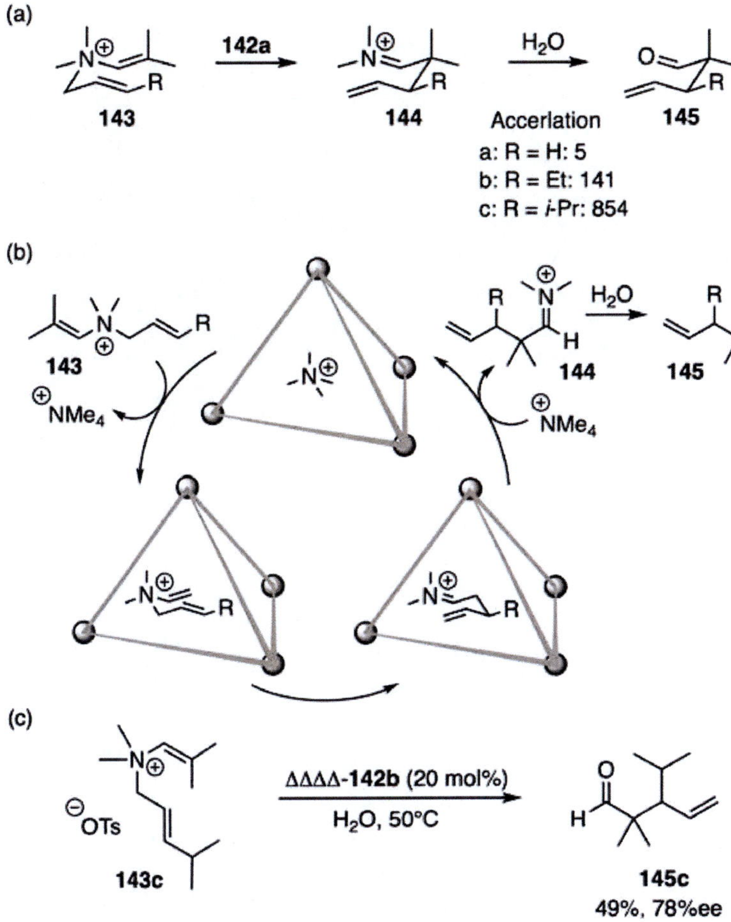

Scheme 4.35 Aza−Cope rearrangement in the tetrahedral supramolecular cage Ga_4L_6 **142**. Reprinted with permission from [1]. Copyright 2018 American Chemical Society

4.4.2.3 Neutral Cages

Resorcin [4] arene **146** is known to form the self-assembling hexamer **147** (**147** ≡ (**146**)$_6$·(H_2O)$_8$) (Fig. 4.17). The structure of hexamer **147** was elucidated by X-ray crystallography, and ^1H NMR studies revealed that eight water molecules are involved in the structure of **146** in $CDCl_3$ solution [48].

The host-catalyzed cyclodehydration-rearrangement of unsaturated tertiary alcohols was reported by Tiefenbacher et al. [49] When cyclopentyl alcohol **148** was added to a solution of hexamer **147** (10 mol%) in $CDCl_3$ and stirred for 4 days, cyclopentene derivative **149** was produced in 81% yield (Scheme 4.36). ^1H NMR studies revealed the encapsulation of alcohol **148** in the cage. The effectiveness of capsule **147** in the cyclization reactions may be due to the stabilization of the cationic intermediates **150−152** as well as the transition state in the cavity.

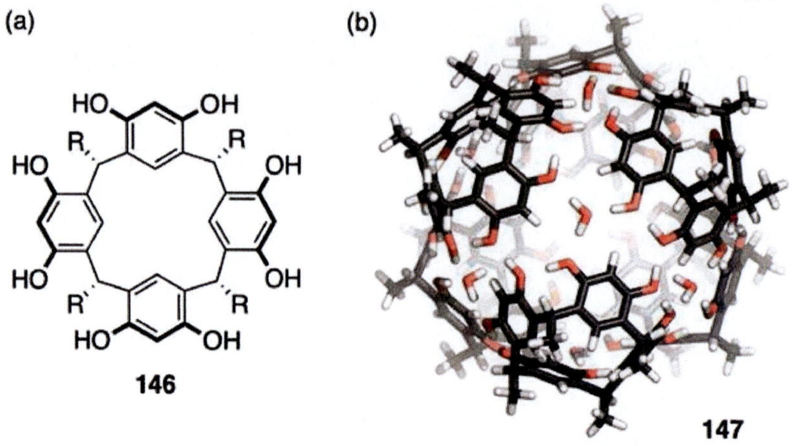

Fig. 4.17 Resorcinarene **146 a** and the self-assembled hexamer **147 b**. From [49] Copyright © 2017 by John Wiley Sons, Inc. Reprinted by permission of John Wiley & Sons, Inc.

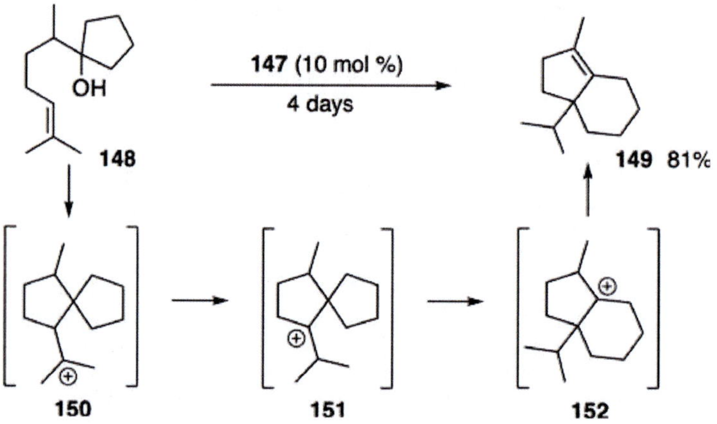

Scheme 4.36 Cyclization of tert-alcohol in a resorcinarene hexamer

4.5 Solid-State Reactions

It is obvious that molecules in crystals are regularly arranged, and it is clear that the pattern is generally governed by various intermolecular interactions such as hydrogen bonding, π–π interactions, and van der Waals forces. Therefore, it is generally difficult to control the arrangement of molecules in crystals. On the other hand, cation–π interactions have been reported to be an effective tool in controlling the arrangement of molecules in a crystalline state [50]. In this section, their use in a variety of solid-state reactions is described.

4.5.1 Photocyclization

4.5.1.1 Azachalcones

It is reported that solid-phase [2+2] photodimerization of 4'-methoxy-4-azachalcone **153** affords *syn*HT dimer **154** and *syn*HH dimer **155** in 20% and 67% yields, respectively (Scheme 4.37). In contrast, irradiation of its hydrochloride salt **153**·HCl yielded only *syn*HT dimer **154** in quantitative yield (Scheme 4.37) [51]. This significant difference in the product selectivities between the reactions of **153** and **153**·HCl is due to differences in the orientation mode of the molecules in these crystals. Figure 4.18 shows the X-ray crystal structures of **153** and **153**·HCl; the molecules of **153** are arranged head-to-head, whereas those of **153**·HCl are arranged head-to-tail. This strongly suggests that cation–π interactions contribute to the molecular arrangement of **153**·HCl.

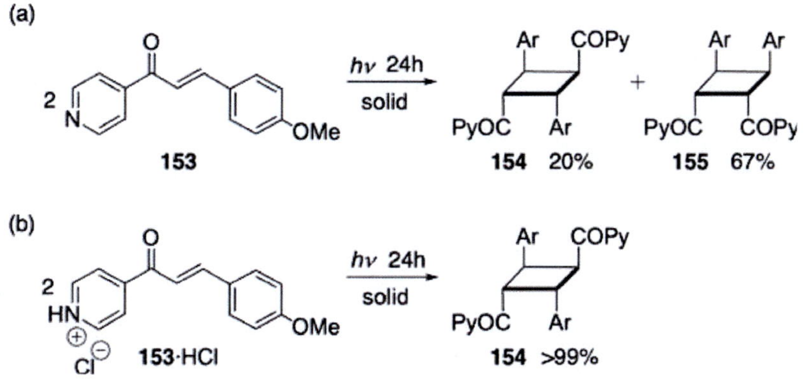

Scheme 4.37 [2+2] Photodimerizations of **a** 4′-methoxy-4-azachalcone and **b** its hydrochloride salt

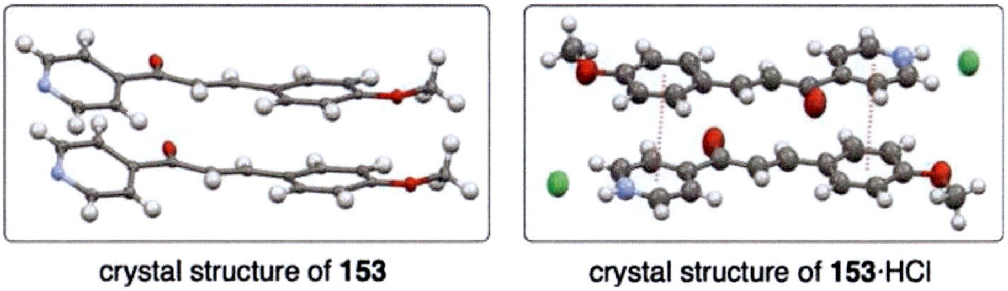

Fig. 4.18 X-ray crystal structures of **153** (left) and **153**·HCl (right). Reprinted with permission from [1]. Copyright 2018 American Chemical Society

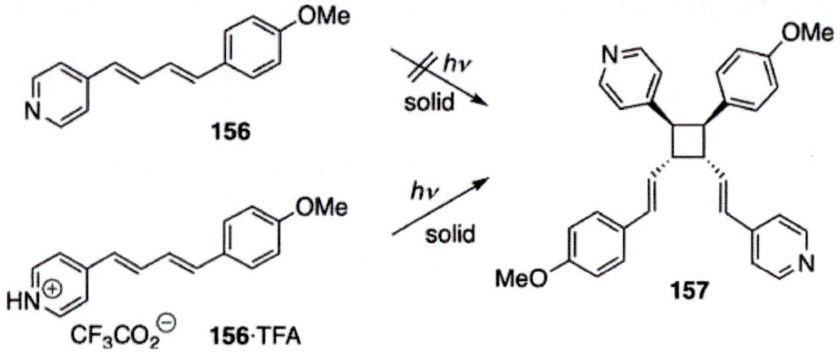

Scheme 4.38 Photodimerization of **156** and **156**·TFA in a solid state

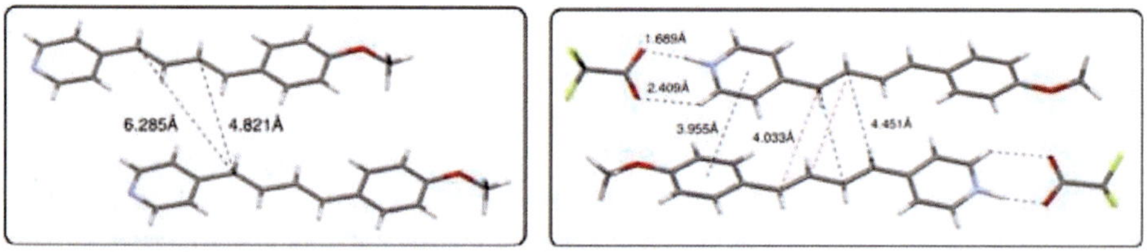

Fig. 4.19 X-ray crystal structures of **156** and **156**·TFA. Reprinted from [53], Copyright 2020, with permission from Elsevier

4.5.1.2 Dienes

It is reported that irradiation of 1-aryl-4-pyridylbutadiene **156** produces *syn*- and *anti*HT dimers when irradiated in the presence of 1 equiv of HCl in methanol solution [52]. When the photochemical reaction was performed in the solid phase, the yield and selectivity were dramatically improved (Scheme 4.38) [53]. Irradiation of the crystals of neutral substrate **156** did not yield a dimer, but irradiation of trifluoroacetate salts **156**·TFA selectively gave *syn*HT dimer **157** in 85% yield.

The crystal structures of **156** and **156**·TFA are shown in Fig. 4.19. The neighboring molecules of **156** are arranged in a head-to-head fashion, and the distance between the double bonds is longer than the Schmidt requirement (<4.2 Å). On the other hand, in the crystal of **156**·TFA, the neighboring molecules are aligned in a head-to-tail (HT) manner and the distance between the double bonds is 4.033 Å, satisfying the Schmidt rule. These structural features of the TFA salt can be explained by the contribution of cation–π interactions.

4.5.1.3 Pyrylium Salts

Novak et al. reported that exposure of a single crystal of styrylpyrylium salt **158** to visible light (>570 nm) led to [2+2] photodimerization, giving *syn*HT dimer **159**

(Scheme 4.39) [54]. This conversion proceeded under the single crystal to single crystal (SCSC) process.

This transformation was applied to the synthesis of two-dimensional polymers, with Schlüter et al. demonstrating the SCSC synthesis of two-dimensional polymers by [2+2] photodimerization of **160** (Scheme 4.40) [55]. Figure 4.20a shows the crystal structure of monomer **160**, in which six styrylpyrylium monomers are arranged in a hexagonal manner in layers that form a column with head-to-tail orientation through intermolecular cation–π interactions. [2+2] Photocyclization of the monomer crystals afforded the two-dimensional polymers with pores as shown in Scheme 4.40 and Fig. 4.20b.

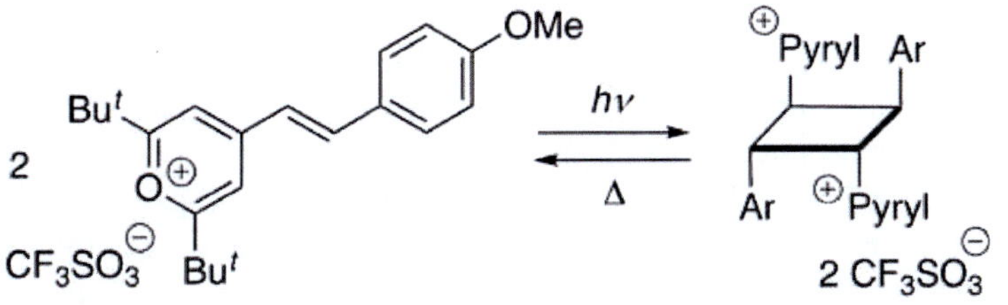

158 **159**

Scheme 4.39 [2+2] Photodimerization of a styrylpyrylium salt

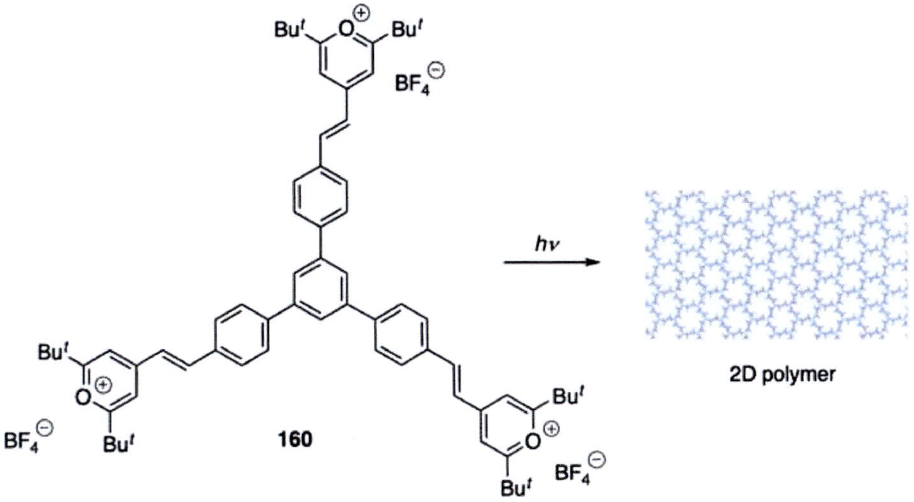

Scheme 4.40 SCSC synthesis of the 2D polymer of a styrylpyrylium derivative by [2+2] photodimerization with the crystal packing diagram of the obtained porous polymer. Reprinted with permission from [55]. Copyright 2017 American Chemical Society

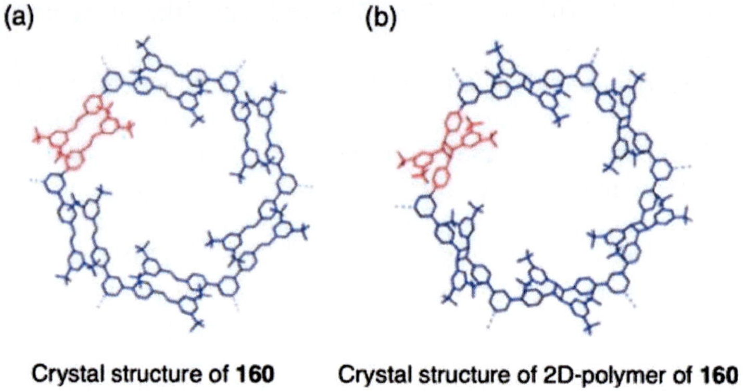

Fig. 4.20 X-ray crystal structures of **160** **a** before irradiation and **b** after irradiation. Reprinted with permission from [55]. Copyright 2017 American Chemical Society

4.5.2 Gas–Solid Reactions

The [2+2] photodimerization of 4'-methoxy-4-azachalcone **153** described in Sect. 4.5.1.1 was applied for the first time to a solid-state cascade reaction. Exposure of crystals of **153** with a head-to-head orientation to HCl gas gave the corresponding HCl salts (**153**·HCl) with a head-to-tail arrangement. Irradiation of the resulting HCl salts yielded *syn*HT dimers in high regio- and stereoselectivities (Scheme 4.41) [56].

Figure 4.21 shows photomicrographs of a single crystal of **153** before and after exposure to HCl gas. After 10 min of exposure, cracks were observed and the initial transparency was almost lost. Irradiation of the resultant **153**·HCl for 24 h showed slight bending of the crystals. PXRD measurements of this crystal were consistent with the simulated pattern obtained from the single crystal of **153**·HCl. This suggests that the crystal obtained by HCl exposure has the same orientation as the single crystal (Fig. 4.22).

^{1}H NMR measurements of the crystals after irradiation showed quantitative formation of *syn*HT dimer **154**, showing that the molecules in the HCl salt are arranged in a head-to-tail fashion (Fig. 4.23). The driving force behind the reorientation from a head-to-head to a head-to-tail manner upon exposure to HCl gas is thought to be the

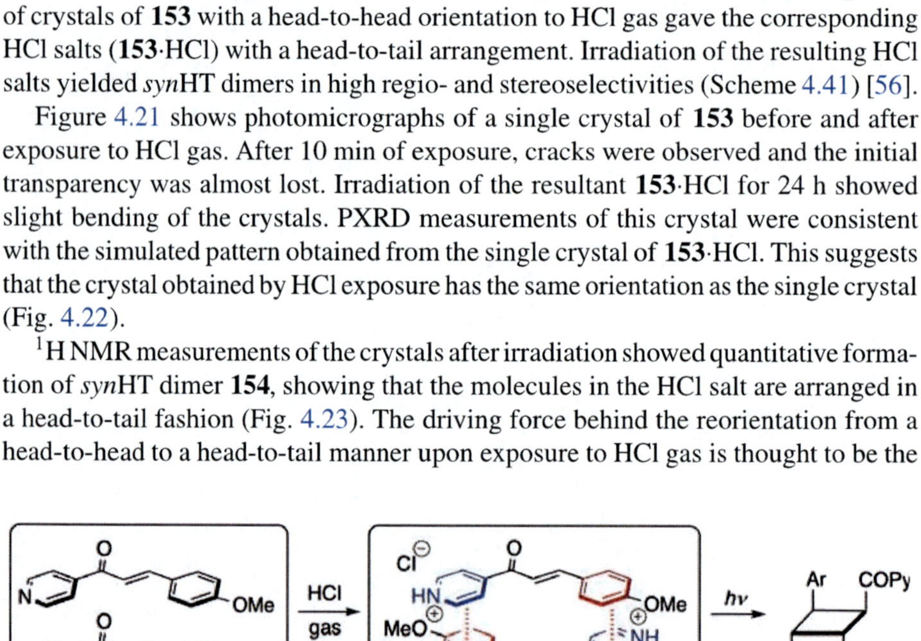

153 **153**·HCl **154** >99%

Scheme 4.41 Solid-state cascade reaction of 4′-methoxy-4-azachalcone **153** to *syn*HT dimer **154**. Reprinted with permission from [1]. Copyright 2018 American Chemical Society

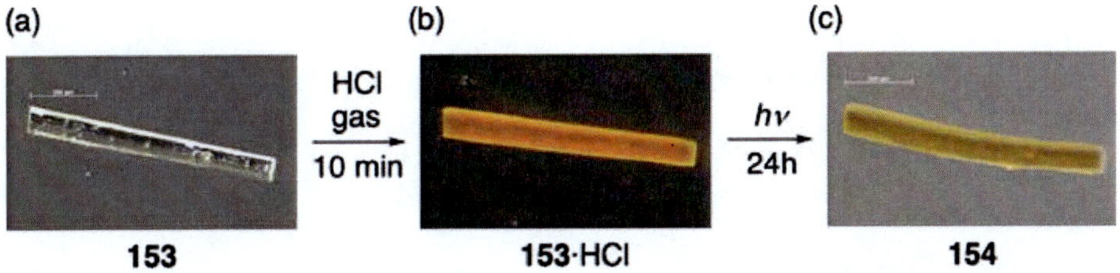

Fig. 4.21 Photomicrographs of a single crystal of **153** **a** before and **b** after exposure to HCl gas for 10 min and **c** after irradiaion of the resultant HCl salt for 24 h. Reprinted with permission from [1]. Copyright 2018 American Chemical Society

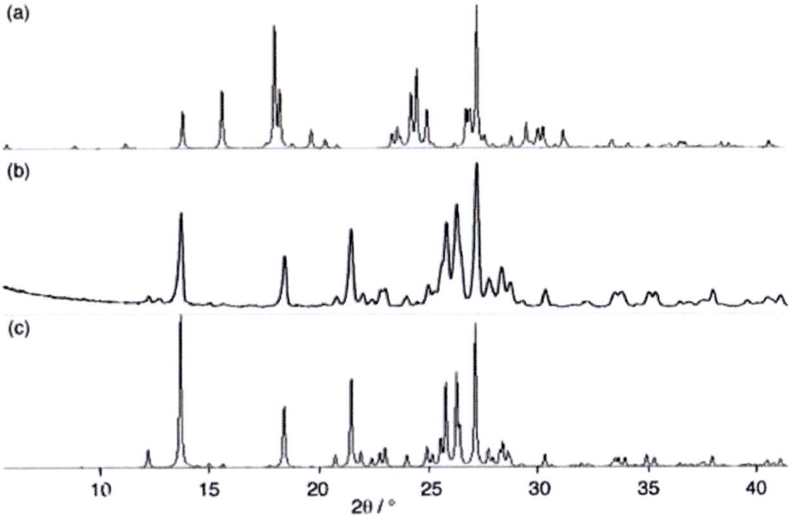

Fig. 4.22 PXRD patterns of **153** **a** before and **b** after exposure to HCl gas for 10 min. **c** Simulated PXRD pattern of **153**·HCl. Reproduced from [56] with permission from the Royal Society of Chemistry

stabilization of the HCl salt by cation–π interactions between the pyridinium and benzene rings.

Although the details of the mechanism by which the orientation change occurs while retaining crystallinity are not clear, the following can be considered as one possible mechanism for this dramatic orientation change involving minimal molecular movement. As shown in Fig. 4.23, when molecules A and B rotate clockwise and A' and B' counterclockwise, cation–π interactions cause the formation of a columns consisting of A and B and A' and B' along the b-axis. By such a mechanism, it is possible that a seemingly large change in orientation could be rearranged by a slight molecular rotation so that the cation–π interaction works.

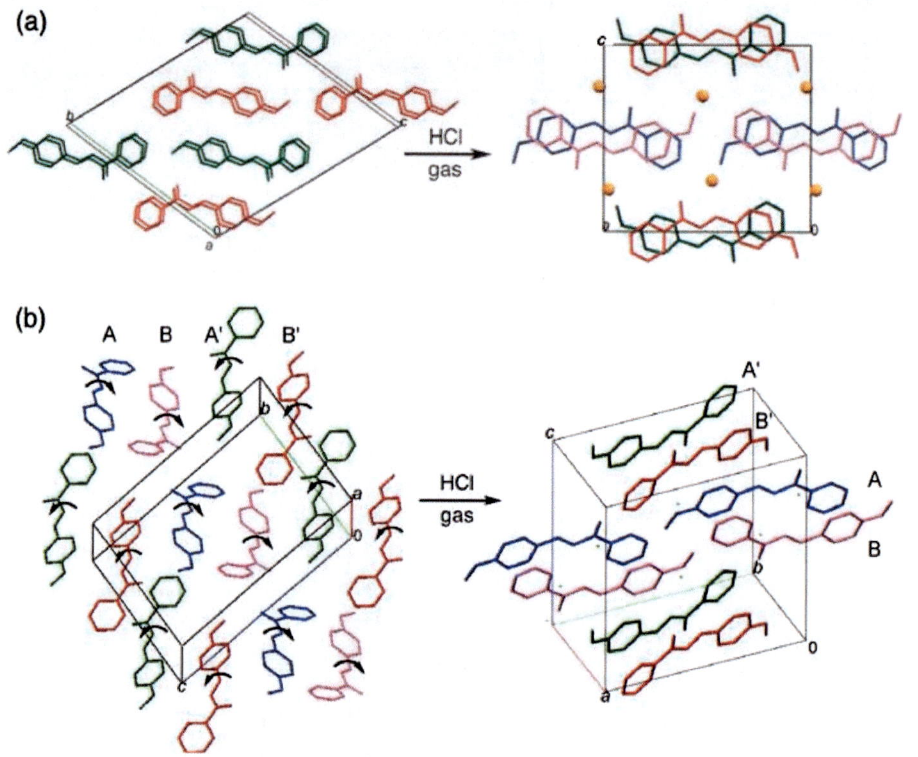

Fig. 4.23 Orientation change of the molecules of **153** on exposure to HCl gas. **a** The crystal packing structures before and after exposure to HCl gas and **b** plausible mechanism for the reorientation of molecules. Reprinted from [50], Copyright 2020, with permission from Elsevier

4.6 Biocatalysis

Cation–π interactions are involved in the three-dimensional structure of proteins and the recognition of substrates. In general, enzymatic reactions generally proceed in a regio- and stereospecific manner, and very attractive from a synthetic perspective. In this section, examples of several reactions using enzymes and designed monoclonal antibodies are presented, in which cation–π interactions are involved in the reaction processes.

4.6.1 Enzymes

It is well known that cation–π interactions play an important role in various enzymatic reactions. Polyene cyclase is an important enzyme in the biosynthesis of various terpenes and steroids. X-ray crystallographic studies of these enzymes in complex

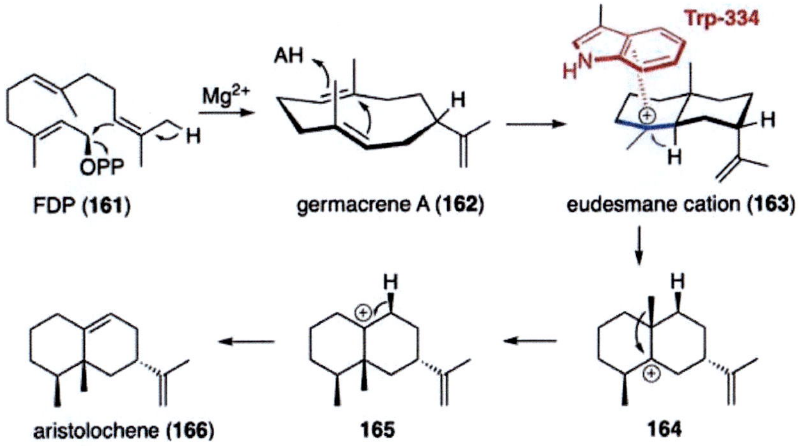

Scheme 4.42 Catalytic cyclization of farnesyl pyrophosphate with aristolochene synthase. Reprinted with permission from [1]. Copyright 2018 American Chemical Society

with their substrates have revealed that aromatic amino acid residues are present in the active site, stabilizing cationic intermediates and transition states.

Scheme 4.42 shows a proposed mechanism of cyclization of farnesyl diphosphate **161** to aristolochene **166** by aristolochene synthase (AS), in which Trp334 plays a pivotal role in the efficient cyclization reaction to produce the bicyclic sesquiterpene, aristolochene **166** [57]. The X-ray crystal structure of AS elucidated that the π-system of Trp334 stabilizes the positive charge of the eudesmane cation **163**. The mutant strain ASW334F showed a 350-fold increase in catalytic efficiency, while 334L and 334V showed a decrease in catalytic efficiency by more than four orders of magnitude, strongly suggesting stabilization of the carbocation by aromatic residues. Substitution of Trp334 with a para-substituted phenylalanine with an electron-withdrawing group resulted in the accumulation of germacrene A **162**, confirming that Trp334 stabilizes the eudesmane cation **163** in the aristolochene synthase.

Farnesyl diphosphate **161** is converted to pentalenene **171** by pentalenene synthase. A plausible mechanism for this conversion is shown in Scheme 4.43, where intramolecular cyclization gives the anti-Markovnikov cation **167** and subsequent 1, 2-hydride shift gives the intermediate **168**. Ring closure of the cation forms the 7-protoilludyl cation **169**, and skeletal rearrangement of the resulting cation **169** yields **170**, whose deprotonation gives pentalenene **171** [58]. Based on the X-ray crystal-structure analysis of pentalenene synthase, the anti-Markovnikov process via **167** and **168** can be explained by stabilization of the intermediate cation by Trp308 through the cation–π interaction. Furthermore, the intermediate cations **169** and **170** are believed to be stabilized by Phe76 and Phe77.

Polycyclization reactions of oxidosqualene catalyzed by oxidosqualene cyclase have attracted much attention, and a number of studies on the underlying mechanism have been reported. For example, Griffin et al. proposed a mechanism for the

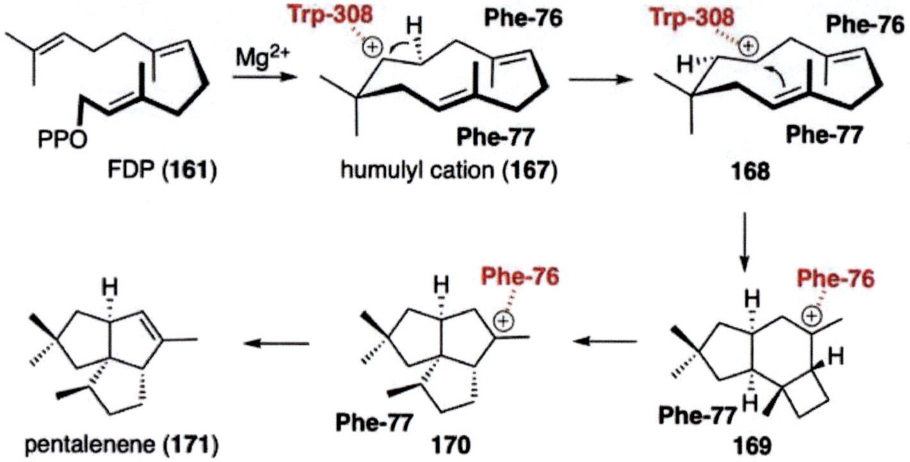

Scheme 4.43 Cyclization of farnesyl pyrophosphate with pentalenene synthase. Reprinted with permission from [1]. Copyright 2018 American Chemical Society

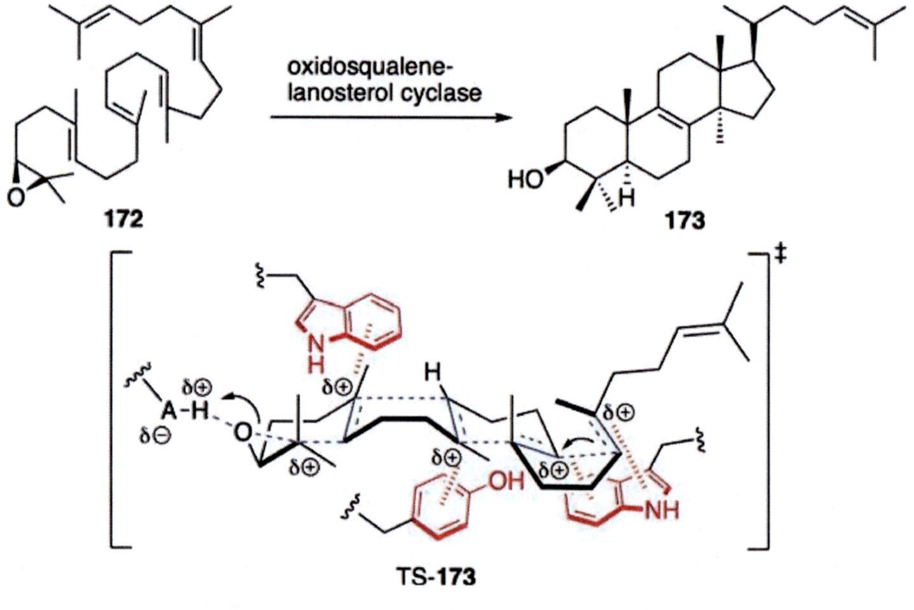

Scheme 4.44 Transformation of (3*S*)-2, 3-oxidosqualene to lanosterol catalyzed by oxidosqualene−lanosterol cyclase with a plausible transition state for the formation of **173**. Reprinted with permission from [1]. Copyright 2018 American Chemical Society

conversion of (3*S*)-2,3-oxodosqualene **172** to lanosterol **173** catalyzed by oxidosqualene–lanosterol cyclase from Saccharomyces cerevisiae [59]. Scheme 4.44 shows the proposed mechanism, where electron-rich tryptophan and tyrosine residues are located in the active site and control the cyclization mode of the substrate by stabilizing the positively charged transition state TS-173 through cation–π interactions.

4.6.2 Catalytic Antibodies

Christianson and coworkers found that antibody HA5-19A4 catalyzed the tandem cationic cyclization of polyene **174** to produce a 1:3:2 mixture of three *trans*-decalin derivatives **175–177** in 50% yield [60]. The enantiomeric excess for **175** and **176** was 53% and for **177** was 80% (Scheme 4.45). As antibody HA5-19A4 was produced against hapten **178**, partially mimicking the chair-chair transition state TS- 175–177 required for cyclization of polyene to *trans*-decalin, this antibody is effective for the cyclization reaction.

The X-ray crystal structure of the Fab fragment of the catalytic antibody complexed with hapten **178** clarified that the aromatic rings of Tyr^{L36} and Trp^{L91} are close to C5 and Tyr^{L36} and Trp^{H103} are close to C9, suggesting the stabilization of C5 and C9 cations through cation–π interactions in the transition state described in Fig. 4.24.

As the catalytic antibody often gives products that are not obtainable in general organic reactions, they are useful for organic synthesis. For example, the intramolecular ring-opening reaction of epoxide **179** generally affords tetrahydrofuran derivative **180** according to Baldwin's law through transition state TS-**180** [61]. In contrast, the reaction using catalytic antibody 26D9, which was made against hapten **182**, yielded the anti-Baldwin product **181** via TS-181 (Scheme 4.46). This suggests that

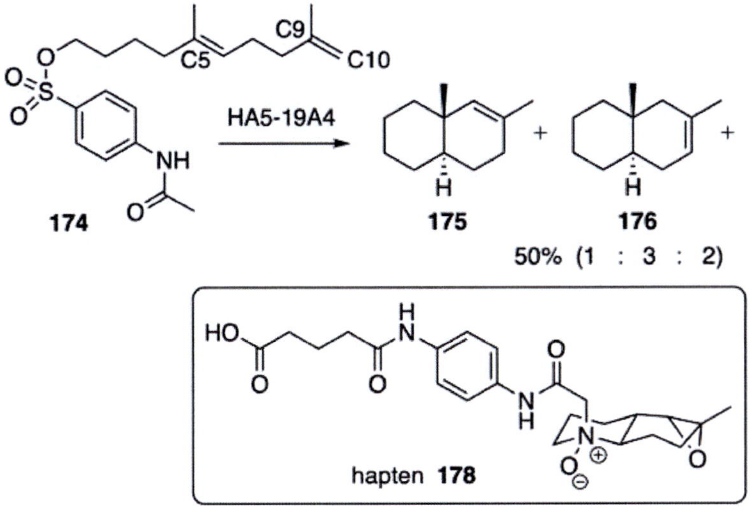

Scheme 4.45 Tandem cationic cyclization of polyene with antibody HA5-19A4. Reprinted with permission from [1]. Copyright 2018 American Chemical Society

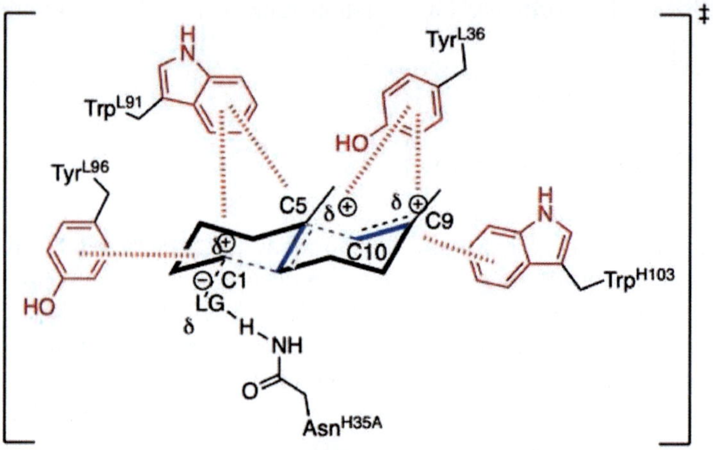

Fig. 4.24 Plausible transition state for the formation of **175**–**177**. Reprinted with permission from [1]. Copyright 2018 American Chemical Society

the cationic nitrogen atom of hapten **182** led the antibody to stabilize the carbocation that develops during the ring-opening reaction.

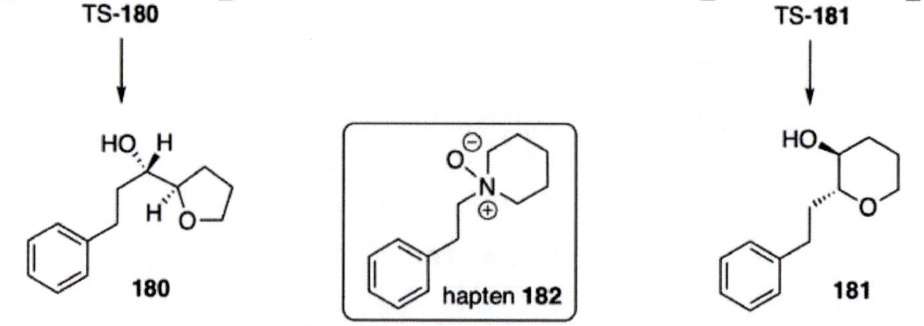

Scheme 4.46 Plausible mechanism for the 5-exo-tet and 6-endo-tet cyclizations of epoxyalcohol. Reprinted with permission from [1]. Copyright 2018 American Chemical Society

References

1. Yamada S (2018) Cation–π interactions in organic synthesis. Chem Rev 118:11353–11432
2. Yamada S, Morita C (2002) Face-selective addition to a cation–π complex of a pyridinium salt: synthesis of chiral 1,4-dihydropyridines. J Am Chem Soc 124:8184–8185
3. Comins DL, LaMunyon DH (1992) Enantiopure 2,3-dihydro-4-pyridones as synthetic intermediates. asymmetric syntheses of the quinolizidine alkaloids (+)-Myrtine, (-)-Lasubine I, and (+)-Subcosine I. J Org Chem 57:5807–5809
4. Singh P, Comins DL, Joseph SP (1994) 4-Methoxy-1-[(1*R*, 4*S*, 5*R*)-8-phenylmenthoxycarbonyl]pyridinium hexachloroantimonate, $C_{23}H_{30}NO_3^+SbCl_6^-$. Acta Cryst C50:25–27
5. Yamada S, Misono T, Iwai Y, Masumizu A, Akiyama Y (2006) New class of pyridine catalyst having a conformation switch system: asymmetric acylation of various *sec*-alcohols. J Org Chem 71:6872–6880
6. Yamada S, Yamashita K (2008) Dynamic kinetic resolution of hemiaminals using a novel DMAP catalyst. Tetrahedron Lett 49:32–35
7. Birman VB, Li X, Jiang H, Uffman EW (2006) Influence of electronic and steric factors on 2,3-dihydroimidazo[1,2-a]pyridine-based enantioselective acylation catalysts. Tetrahedron 62:285–294
8. Matsumoto M, Lee SJ, Gagne MR, Waters ML (2014) Cross-strand histidine-aromatic interactions enhance acyl-transfer rates in β-hairpin peptide catalysts. Org Biomol Chem 12:8711–8718
9. West TH, Walden DM, Taylor JE, Brueckner AC, Johnston RC, Cheong PH, Lloyd-Jones GC, Smith AD (2017) Catalytic enantioselective [2, 3]-rearrangements of allylic ammonium ylides: a mechanistic and computational study. J Am Chem Soc 139:4366–4375
10. Williams JLR, Carlson JM, Reynolds GA, Adel RE (1963) Photodimers of 4'-substituted 2-styrylpyridines. J Org Chem 28:1317–1320
11. Yamada S, Uematsu N, Yamashita K (2007) Role of cation–π interactions in the photodimerization of *trans*-4-styrylpyridines. J Am Chem Soc 129:12100–12101
12. Yamada S, Kawamura C (2012) [4+4] Photodimerization of azaanthracenes in both solution and solid phase controlled by cation–π interactions. Org Lett 14:1572–1575
13. Bolduc P, Jacques A, Collins SK (2010) Efficient macro-cyclization achieved via conformational control using intermolecular noncovalent π-Cation/Arene interactions. J Am Chem Soc 132:12790–12791
14. Barnes JC, Juricek M, Strutt NL, Frasconi M, Sampath S, Giesener MA, McGrier PL, Bruns CJ, Stern CL, Sarjeant AA, Stoddart JF (2013) ExBox: a polycyclic aromatic hydrocarbon scavenger. J Am Chem Soc 135:183–192
15. Kumar R, Saima SA, Andhare NH, Richa SAK (2015) Thiol-ene "click" reaction triggered by neutral ionic liquid: the "ambiphilic" character of [hmim]Br in the regioselective nucleophilic hydrothiolation. Angew Chem Int Ed 54:828–832
16. Feng L, Ye R, Yuan T, Zhang X, Lu G-P, Zhou B (2019) A concerted addition mechanism in [Hmim]Br-triggered thiol–ene reactions: a typical "ionic liquid effect" revealed by DFT and experimental studies. New J Chem 43:5752–5758
17. Narumi T, Tsuzuki S, Tamamura H (2014) Imidazolium salt-catalyzed Friedel−Crafts-type conjugate addition of indoles: analysis of indole/imidazolium complex by high level ab initio calculations. Asian J Org Chem 3:497–503
18. Comins DL, Thakker PM, Matthew FB, Badawi MM (1997) Chiral auxiliary mediated pictet-spengler reactions: asymmetric syntheses of (-)-Laudanosine, (+)-Glaucine and (-)-Xylopinine. Tetrahedron 53:16327–16340
19. Scharf MJ, List B (2022) A catalytic asymmetric pictet-spengler platform as a biomimetic diversification strategy toward naturally occurring alkaloids. J Am Chem Soc 144:15451-15456
20. Ahrendt KA, Borths CJ, MacMillan DWC (2000) New strategies for organic catalysis: the first highly enantioselective organocatalytic Diels-Alder reaction. J Am Chem Soc 122:4243–4244

21. Mori Y, Yamada S (2012) Contribution of cation–π interactions in iminium catalysis. Molecules 17:2161–2168
22. Austin JF, MacMillan DWC (2002) Enantioselective organocatalytic indole alkylations: Design of a new and highly effective chiral amine for iminium catalysis. J Am Chem Soc 124:1172–1173
23. Groselj U, Schweizer WB, Ebert M-O, Seebach D (2009) 5-Benzyl-3-methylimidazolidin-4-one-derived reactive intermediates of organocatalysis—a comforting resemblance of X-Ray, NMR, and DFT solid-phase, liquid-phase, and gas-phase structures. Helv Chim Acta 92:1–13
24. Knowles RR, Lin S, Jacobsen EN (2010) Enantioselective thiourea-catalyzed cationic polycyclizations. J Am Chem Soc 132:5030–5032
25. Banik SM, Levina A, Hyde AM, Jacobsen EN (2017) Lewis acid enhancement by hydrogen-bond donors for asymmetric catalysis. Science 358:761–764
26. Yamamoto-Katayama S, Ariyoshi M, Ishihara K, Hirano T, Jingami H, Morikawa K (2002) Crystallographic studies on human BST-1/CD157 with ADP-ribosyl cyclase and NAD glycohydrolase asctivities. J Mol Biol 316:711–723
27. Yamada S, Gu Q-M, Sih CJ (1994) Cyclic ADP-ribose via stereoselective cyclization of β-NAD. J Am Chem Soc 116:10787–10788
28. Denis C, Dubois MAJ, Voisin-Chiret AS, Bureau R, Choi C, Mousseau JJ, Bull JA (2019) Synthesis of 3,3-diarylazetidines by calcium(II)-catalyzed Friedel–Crafts reaction of azetidinols with unexpected Cbz enhanced reactivity. Org Lett 21:300–304
29. Yamato T, Tokuhisa K, Tsuzuki H (2000) Medium-sized cyclophanes. Part 51. Acylation of [2.2]metaparacyclophanes: through-space electronic interactions between two benzene rings. Can J Chem 78:238–247
30. Yamada S, Iwaoka A, Fujita Y, Tsuzuki S (2013) Tetraalkylammonium-templated stereoselective Norrish-Yang cyclization. Org Lett 15:5994–5997
31. Johnston CP, Kothari A, Sergeieva T, Okovytyy SI, Jackson KE, Paton RS, Smith MD (2015) Catalytic enantioselective synthesis of indanes by a cation-directed 5-endo-trig cyclization. Nat Chem 7:171–177
32. Hiroya K, Jouka R, Kameda M, Yasuhara A, Sakamoto T (2001) Cyclization reactions of 2-Alkylbenzyl alcohol and 2-Alkynylbenzyl-amine derivatives promoted by Tetrabutylammonium Fluoride. Tetrahedron 57:9697–9710
33. Fujii A, Choi J-C, Fujita K-I (2017) Quaternary ammonium salt-catalyzed carboxylative cyclization of propargylic amines with CO_2. Tetrahedron Lett 58:4483–4486
34. Zhao Y, Zhang Z, Liu X, Wang Z, Cao Z, Tian L, Yue M, You J (2019) TBAF-catalyzed O-Nucleophilic cyclization of enaminones: a process for the synthesis of dihydroisobenzofuran derivatives. J Org Chem 84:1379–1386
35. Maity P, Lepore SD (2011) Catalytic synthesis of nonracemic azaproline derivatives by cyclization of β-Alkynyl hydrazines under kinetic Rresolution conditions. Angew Chem Int Ed 50:8338–8341
36. Nagy E, Germain E, Cosme P, Maity P Terentis AC, Lepore SD (2016) Ammonium catalyzed cyclitive additions: evidence for a cation–π interaction with alkynes. Chem Commun 52:2311-2313
37. Zhang WB, Lin G, Shao W-B, Gong J-X, Yang Z (2015) Total synthesis of maoecrystal V. Chem Asian J 10:903–909
38. Cousins GRL, Furlan RLE, Ng Y-F, Redman JE, Sanders JKM (2001) Identification and isolation of a receptor for N-methyl alkylammonium salts: molecular amplification in a pseudo-peptide dynamic combinatorial library. Angew Chem Int Ed 40:423–428
39. Posner GH, Lentz CM (1979) A directing effect of neighboring aromatic groups on the regiochemistry of formation and on the stereochemistry of alkylation and bromination of ketone lithium enolates: evidence for lithium-arene π coordination and for a dramatic effect of even small amounts of copper(1) in controlling stereochemistry and in limiting polyalkylation. J Am Chem Soc 101:934–946
40. Sini G, Tessier A, Pytkowicz J, Brigaud T (2008) Fluorine··· and π···alkali metal interactions control in the stereoselective amide enolate alkylation with fluorinated oxazolidines (Fox) as a chiral auxiliary: an experimental and theoretical study. Chem Eur J 14:3363–3370

41. Peng X, Tong BMK, Hirao H, Chiba S (2014) Inorganic-base-mediated hydroamination of alkenyl oximes for the synthesis of cyclic nitrones. Angew Chem Int Ed 53:1959–1962
42. Stauffer DA, Barrans RE Jr, Dougherty DA (1990) Biomimetic catalysis of an S_N2 reaction resulting from a novel form of transition-state stabilization. Angew Chem Int Ed Engl 29:915–918
43. Richeter S, Rebek J Jr (2004) Catalysis by a synthetic receptor sealed at one end and functionalized at the other. J Am Chem Soc 126:16280–16281
44. Yoshizawa M, Takeyama Y, Kusukawa T, Fujita M (2002) Cavity-directed, highly stereoselective [2+2] photodimerization of olefins within self-assembled coordination cages. Angew Chem Int Ed 41:1347–1349
45. Nishioka Y, Yamaguchi T, Kawano M, Fujita M (2008) Asymmetric [2+2] olefin cross photoaddition in a self-assembled host with remote chiral auxiliaries. J Am Chem Soc 130:8160–8161
46. Fiedler D, van Halbeek H, Bergman RG, Raymond KN (2006) Supramolecular catalysis of unimolecular rearrangements: substrate scope and mechanistic insights. J Am Chem Soc 128:10240–10252
47. Brown CJ, Bergman RG, Raymond KN (2009) Enantioselective catalysis of the Aza-cope rearrangement by a chiral supramolecular assembly. J Am Chem Soc 131:17530–17531
48. MacGillivray LR, Atwood JL (1997) A chiral spherical molecular assembly held together by 60 hydrogen bonds. Nature 389:469–472
49. Catti L, Pöthig A, Tiefenbacher K (2017) Host-catalyzed cyclodehydration-rearrangement cascade reaction of unsaturated tertiary alcohols. Adv Synth Catal 359:1331–1338
50. Yamada S (2020) Cation–π interactions in organic crystals. Coord Chem Rev 415:213301
51. Yamada S, Tokugawa Y (2009) Cation–π controlled solid-state photodimerization of 4-azachalcones. J Am Chem Soc 131:2098–2099
52. Yamada S, Azuma Y, Aya K (2014) [2+2] Photodimerization of 1-aryl-4-pyridylbutadienes through cation–π interactions. Tetrahedron Lett 55:2801–2804
53. Yamada S, Aya K (2020) Solid-state [2+2] photodimerization of 1-Aryl-4-pyridylbutadienes in cation–π-controlled crystals. Chem-Asian J 15:581–584
54. Novak K, Enkelmann V, Köhler W, Wegner G, Wagener KB (1994) Homogeneous photodimerization and thermal back reaction of a styrylpyrylium triflate. Mol Cryst Liq Cryst Sci Technol Sect A 242:1–8
55. Lange RZ, Hofer G, Weber T, Schlüter AD (2017) A two-dimensional polymer synthesized through topochemical [2+2]-cycloaddition on the multigram scale. J Am Chem Soc 139:2053–2059
56. Yamada S, Tokugawa Y, Nojiri Y, Takamori E (2012) Cascade reactions in crystals through cation–π-controlled reorientation on exposure to HCl gas. Chem Commun 48:1763–1765
57. Faraldos JA, Antonczak AK, Gonzalez V, Fullerton R, Tippmann EM, Allemann RK (2011) Probing Eudesmane cation–π interactions in catalysis by aristolochene synthase with non-canonical amino acids. J Am Chem Soc 133:13906–13909
58. Seemann M, Zhai G, de Kraker J-W, Paschall CM, Christianson DW, Cane DE (2002) Pentalenene synthase: analysis of active site residues by site-directed mutagenesis. J Am Chem Soc 124:7681–7689
59. Shi Z, Buntel CJ, Griffin JH (1994) Isolation and characterization of the gene encoding 2,3-oxidosqualene-lanosterol cyclase from saccharomyces cerevisae. Proc Natl Acad Sci USA 91:7370–7374
60. Paschall CM, Hasserodt J, Jones T, Lerner RA, Janda KD, Christianson DW (1999) Convergence of catalytic antibody and terpene cyclase mechanisms: polyene cyclization directed by carbocation–π interactions. Angew Chem Int Ed 38:1743–1747
61. Janda KD, Shevlin CG, Lerner RA (1993) Antibody catalysis of a disfavored chemical transformation. Science 259:490–493

Chapter 5
Materials Science

5.1 Polymers

5.1.1 Supramolecular Polymers and Oligomers

5.1.1.1 Helical Polymers

Freire et al. have shown that polyphenylacetylene (Poly-**A**) (Fig.5.1), containing (*R*)-α-methoxyphenylacetic acid (MPA) as a chiral pendant, is in a 1:1 equilibrium between the left (*ap*) and right (*sp*) helical conformations in solution [1]. In addition, they found that this equilibrium shifts in the one helical direction upon the addition of metal salts, which is thought to be due to the formation of helical polymer-metal complexes (HPMCs).

When various $LiClO_4$ solutions were added to the $CHCl_3$ solution of poly-**A**, CD monitoring clarified that the *ap* conformation of the pendant was stabilized in all cases (Fig. 5.2a) and a left-handed helix was induced (negative CD, $g_{384\ nm}<0$) (Fig. 5.2b). In particular, a large cotton effect was observed when THF solvent was used. ^{7}Li NMR, ^{1}H NMR, and IR studies of the poly-**A**/Li^+ solution indicate the presence of cation–π interactions in the structure of the generated HPMC, as shown in Fig. 5.2b [2]. Similar CD curves were observed in DMSO and MeOH solutions, suggesting that the Li^+-π interaction is sufficiently strong in this system. The ternary complex, the M^+-π–π motif, is thought to be responsible for the generation of the helicity through the Li^+-π interaction.

As introduced in Chap. 2, theoretical calculations predict that cooperation among cation–π and π–π interactions leads to the enhancement of each interaction through synergistic effects, where the π–π interaction energy in the presence of cations is enhanced by about 2–5 kcal mol^{-1}. As shown in Fig. 5.3, the cation-$(\pi)_n$-π interaction reinforces the π–π interaction by spreading the cationic charge between the stacked aromatic rings, leading to helicity in the polymer.

Abe et al. reported pH-adjustable sugar recognition using poly(2,6-diethynylpyridine) (**1**) with dialkylamino groups (Fig. 5.4) [3]. The addition of TFA to

S. Yamada, *The Cation-π Interaction*,
https://doi.org/10.1007/978-981-19-7335-2_5

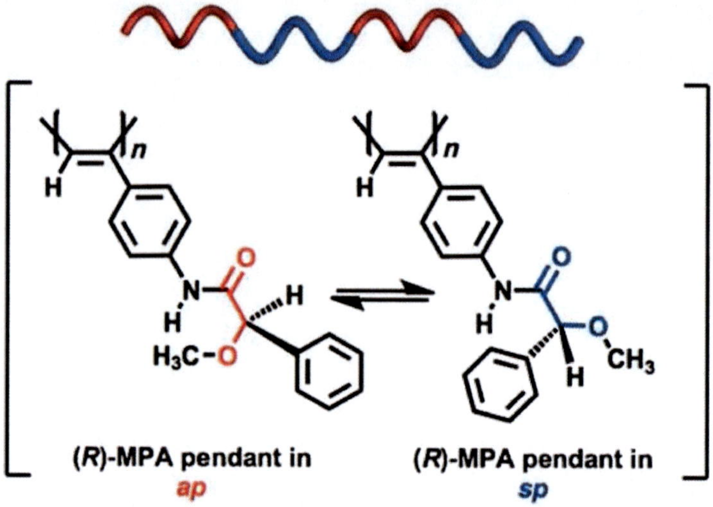

Fig. 5.1 Equilibrium between two helicities induced by the conformation of the pendants. Reproduced from Ref. [2] with permission from the Royal Society of Chemistry

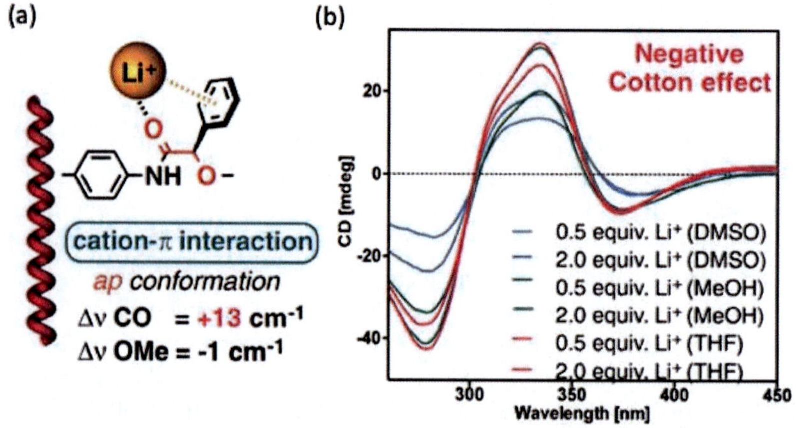

Fig. 5.2 **a** *ap* Conformation stabilized by cation–π interactions and determined by IR. **b** CD spectra of poly-**A**/Li^+ complexes; Li^+ equiv. per mru equiv. are indicated; cosolvents are shown in parentheses; [poly-**A**] = 0.3 mg mL^{-1} in $CHCl_3$, [$LiClO_4$] = 10 mg mL^{-1} in DMSO, MeOH, and THF. From [1] Copyright © 2011 by John Wiley Sons, Inc. Reprinted by permission of John Wiley & Sons, Inc.

a CH_2Cl_2 solution of **1** and sugar led to the formation of a helical complex and resulted in induced circular dichroism (ICD) (Fig. 5.5a). The ICD gradually increased until the amount of acid reached about 50 mol%, at which point the chiral helical structure is generated by the incorporation of D-Glc (Fig. 5.5). These results are attributed to the semiprotonation of the pyridine rings, which stabilized the helical structure consisting of cisoidal coordination through pyridine–pyridinium interactions (Fig. 5.5b).

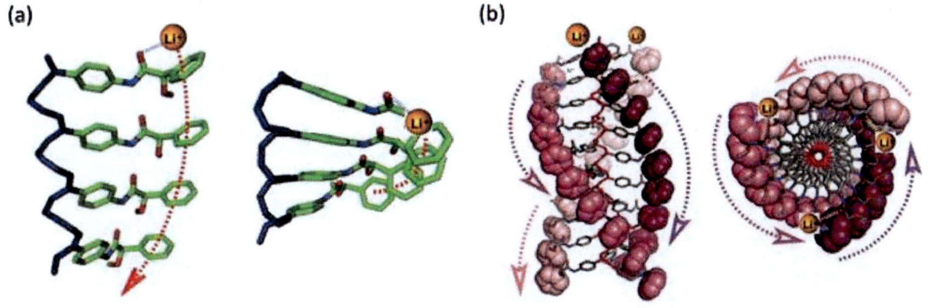

Fig. 5.3 **a** Two views showing the favorable structure for the formation of cation–$(\pi)_n$–π interactions (red broken arrows) in poly-**A**/M$^+$. For the sake of clarity, only one crest of the polymer is shown. **b** Conceptual representation (side and frontal views) indicating that cation–$(\pi)_n$–π interactions can results in the complete 3/1 helical structure of poly-**A**. MPA phenyl groups are shown as space-filling models. From [1] Copyright © 2011 by John Wiley Sons, Inc. Reprinted by permission of John Wiley & Sons, Inc.

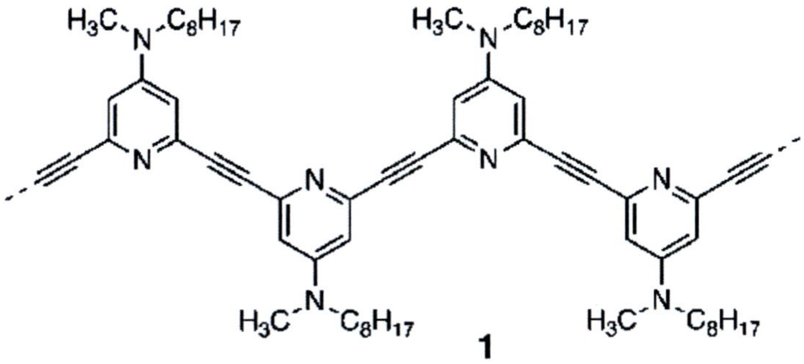

Fig. 5.4 Structure of poly(2,6-diethynylpyridine) **1**

Further addition of the acid suppressed the ICD (Fig. 5.5a). This suggests that when the amount of TFA exceeded 50 mol%, the helical structure was lost due to repulsion between the cations, leading to an inability to recognize D-Glc.

5.1.1.2 Reactivity of Helical Oligomers

Heemstra and Moore reported that phenylene ethynylene (PE) oligomers form a helical structure having hydrophobic cavities [4]. The rate of N-methylation of PE oligomers was dependent on the length of the oligomer. In the N-methylation of oligomers **2a–2c**, oligomer **2b** with 13 monomer units reacted 400 times faster than did oligomer **2a** with 3 monomer units (Fig. 5.6). The pyridine ring of **2b** is sandwiched between the two phenyl rings, which stabilizes the transition state TS-**3** through a pyridinium-π interaction. On the other hand, **2a** is not long enough to benefit from such stabilization and, as a result, the large difference in the reaction rate

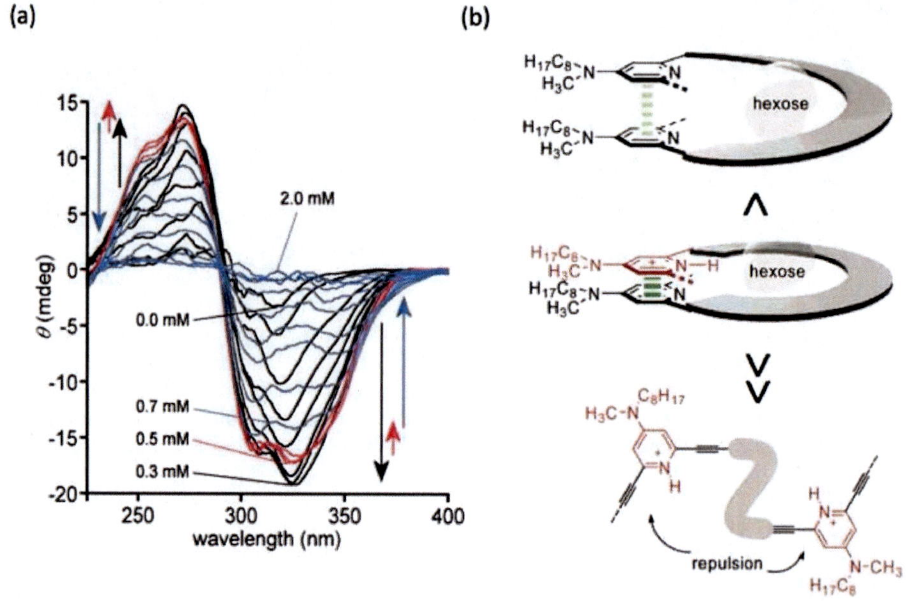

Fig. 5.5 **a** Change in the CD spectrum of the complex between **1** and β-D-Glc on titration with TFA **b** with schematic helical and extended structures. Conditions: [**1**] = 1.0 mM, [β-D-Glc]) 2.5 mM, [TFA] = 0−0.3 mM (black), 0.4−0.6 mM (red), 0.7−2.0 mM (blue), CH_2Cl_2, 26 °C. Light path length 1 mm. Reprinted with permission from [3]. Copyright 2005 American Chemical Society

between them was observed. The methylation rate of **2c** significantly depends on the methylation reagents: The methylation rate with 3-pentylmethylsulfonic acid is 1600 times greater than that of **2a** [5]. On the other hand, when 1-adamantylmethylsulfonic acid is used, the methylation rate is 8 times higher, suggesting that the oligomers form cavities that play a role in sieving reactivity.

Dolain et al. investigated the folding effect of oligoamides on the acceleration of the N-oxidation reactions of **4a** and **4b** at peripheral reactive sites (Fig. 5.7a) [7]. Pentamer **4a** and heptamer **4b** consist of 2,6-diaminopyridine and 2,6-pyridinedicarboxylic acid. The oligoamides fold into stable single helices, facilitating the quaternization of the pyridine moiety. The rate of their N-oxidation with *m*-CPBA to the corresponding N-oxides **5** is much faster than that of the monomer and dimer. The initial reaction rate of **4b** is at least 400 times faster than the oxidation of the dimer derived from its two terminal units. Figure 5.7b shows the crystal structure of the heptamer **4b** with a helical folding structure. One of the reasons for this rate acceleration can be attributed to the stabilization of the N-oxidation transition state by cation–π interactions.

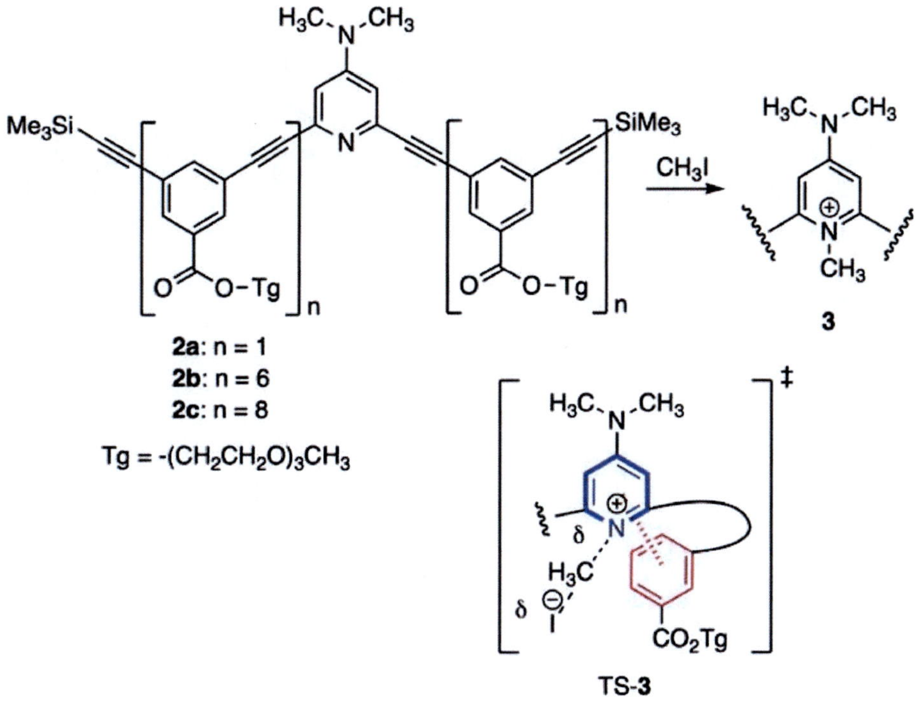

Fig. 5.6 Methylation of the pyridine-containing oligomers **2a–2c**. Reprinted with permission from [6]. Copyright 2018 American Chemical Society

5.1.1.3 Self-Assembling Polymers

Supramolecular polymers combine the useful functions of conventional polymers with dynamic properties based on their formation by non-covalent bonding. To date, a variety of non-covalent intermolecular forces, such as hydrogen bonding and coordination bonds, have been used to generate supramolecular polymers. Aydogan and Sessler reported a new class of supramolecular polymers by the self-assembly of imidazolium-functionalized calix[4]pyrrole [8]. The calix[4]pyrroles (**6**) possessing an imidazolium bromide moiety assemble to produce a supramolecular polymer in which the imidazolium cation is bound to the calix pyrrole core and the bromide anion generates hydrogen bonds with the pyrrole hydrogen outside the ring (Fig. 5.8). The ^{1}H-NMR, 2D-DOSY, and 2D-NOESY spectra confirm that the imidazolium subunit is incorporated into the electron-rich cup of the calix[4]pyrrole.

To gain further evidence for the formation of supramolecular aggregates from **6**, viscosity measurements were performed in $CHCl_3$/DMSO (95/5). As shown in Fig. 5.9, at low concentrations, the slope of the curve is near unity, indicating a linear relationship between specific viscosity and concentration. When the concentration exceeded the critical polymerization concentration (about 12.5 mM in the case of CPC), the viscosity increased rapidly (slope = 3.65), whereas the control experiment using a mixture of 1-ethoxycarbonylcalix[4]pyrrole and imidazolium

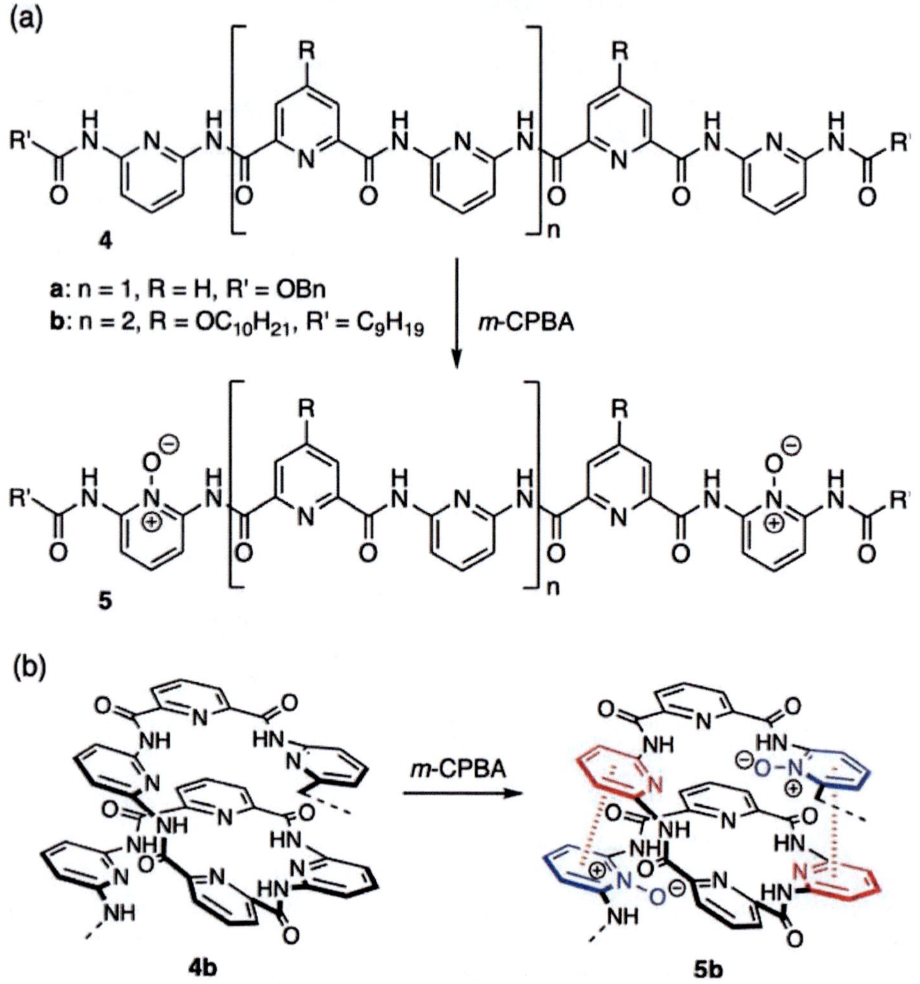

Fig. 5.7 **a** Structure of oligomeric strands **4a** and **4b**. **b** Schematic representation of the stack of four pyridine and pyridine N-oxide rings. Reprinted with permission from [6]. Copyright 2018 American Chemical Society

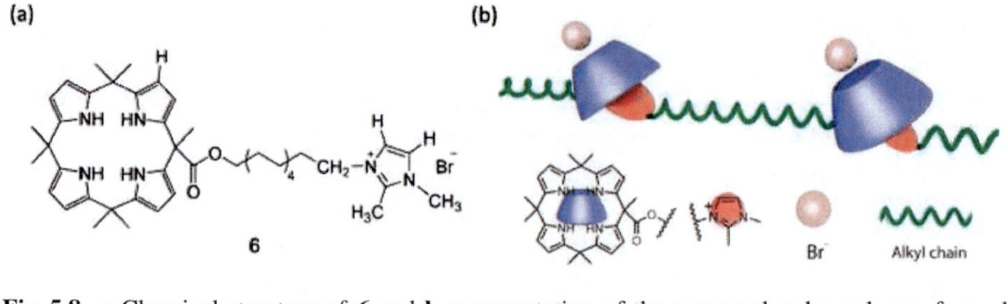

Fig. 5.8 **a** Chemical structure of **6** and **b** representation of the supramolecular polymer formed via the self-assembly of **6**. Reproduced from Ref. [8] with permission from the Royal Society of Chemistry

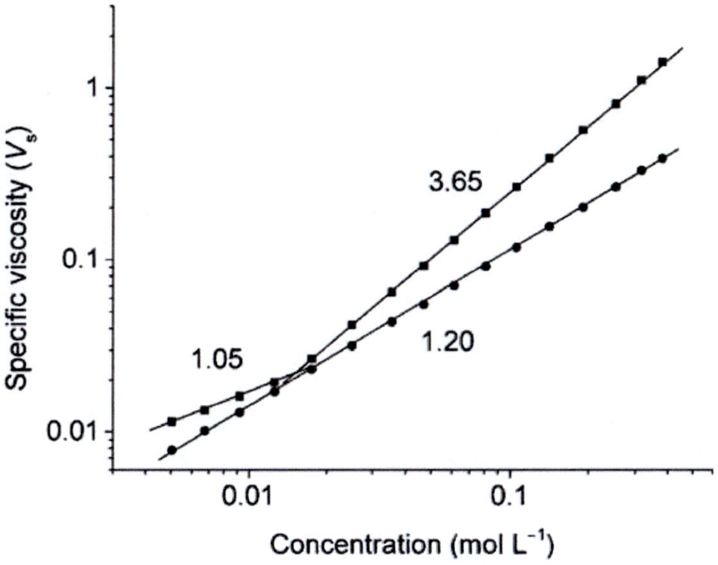

Fig. 5.9 Specific viscosity (303 K) of **6** (black square) and a mixture of calix[4]pyrrole and imidazolium derivatives (black circle) plotted against the monomer concentration. Reproduced from Ref. [8] with permission from the Royal Society of Chemistry

derivative shows a relatively low viscosity (slope = 1.20) as it does not lead to the formation of supramolecular polymers.

Yuvayapan and Aydogan synthesized the carboxylic acid-tethered calix[4]pyrroles **7a** and **7b** with tetrabutylammonium (TBA) and cetyltrimethylammonium (CTA) as cations, respectively [9], which are highly soluble in organic solvents (Fig. 5.10). The supramolecular polymers of **7** exhibited reversible pH-responsiveness; the viscosity in $CHCl_3$ solution decreased with increases in the concentration of TfOH as the addition of the acid breaks the hydrogen bond between the calix[4]pyrrole core and the carboxylate unit, indicating that the supramolecular polymerization of **7** can be controlled by changing the pH of the solution.

A comparison of **7a** and **7b** clarified the fact that CTA enhances the solubility and other properties of supramolecular polymers such as viscosity, drawing fibers, and gel formation. These features can be attributed to the cation–π interaction between the calix[4]pyrrole and the trimethylammonium moiety of CTA as well as the hydrophobic interactions between the long alkyl chains of CTA in the case of **7b**.

5.1.1.4 Supramolecular Thermosetting Resins

Supramolecular chemistry has provided guidance for the design of intelligent and versatile soft materials, but it is not applicable to hard materials. Chang et al. developed a high-performance supramolecular thermosetting resin based on cation–π

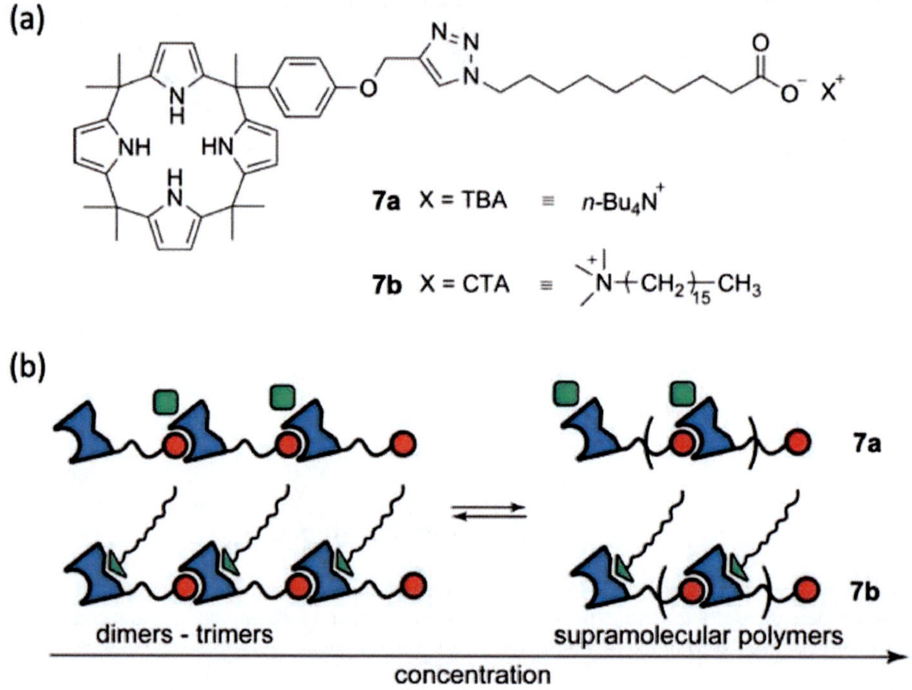

Fig. 5.10 **a** Structures of carboxylate-functionalized calix[4]pyrroles (**7**) and **b** schematic representation of the counter cation-dependent supramolecular polymerization of **7a** and **7b**. From [9] Copyright © 2019 by John Wiley Sons, Inc. Reprinted by permission of John Wiley & Sons, Inc.

interactions [10]. They synthesized sulfonated poly(aryl indole ketone)s (SPAIKs) by random copolymerization, which introduced indole units into the polymer chain as shown in the Fig. 5.11a. Polymer films were prepared from the dimethyl sulfoxide (DMSO) solution of this polymer by solvent casting. These supramolecular thermosets are insoluble solid polymers with excellent mechanical properties even at temperatures above 300 °C. On the other hand, treatment with aqueous solutions of different pH allows for local and reversible cation–π interactions (Fig. 5.11b), which imparts various mechanical properties such as recyclability, repairability, adhesion, and cross-linking to these thermosets. The equilibrium conformation of the SPAIKs studied by quantum dynamics (QD) and molecular dynamics (MD) simulations predicted that the interaction energy for a distance of 3.16 Å between the K^+ and indole is -21.4 kcal mol^{-1} (Fig. 5.12). These results suggest that the change in macroscopic properties is due to the formation of cross-links by cation (K^+)-π interaction.

In the case of the SPAIKs described above, a large amount of indole moieties need to be introduced into the main chain, which greatly increases the production cost of the polymer. To overcome this shortcoming, they prepared a new polymer (SPIN) that contains only a small amount of indole moieties in the polymer side chain (Fig. 5.13a). 5,5'-Sodium carbonylbis(2-fluorobenzenesulfonate) (SDF) and

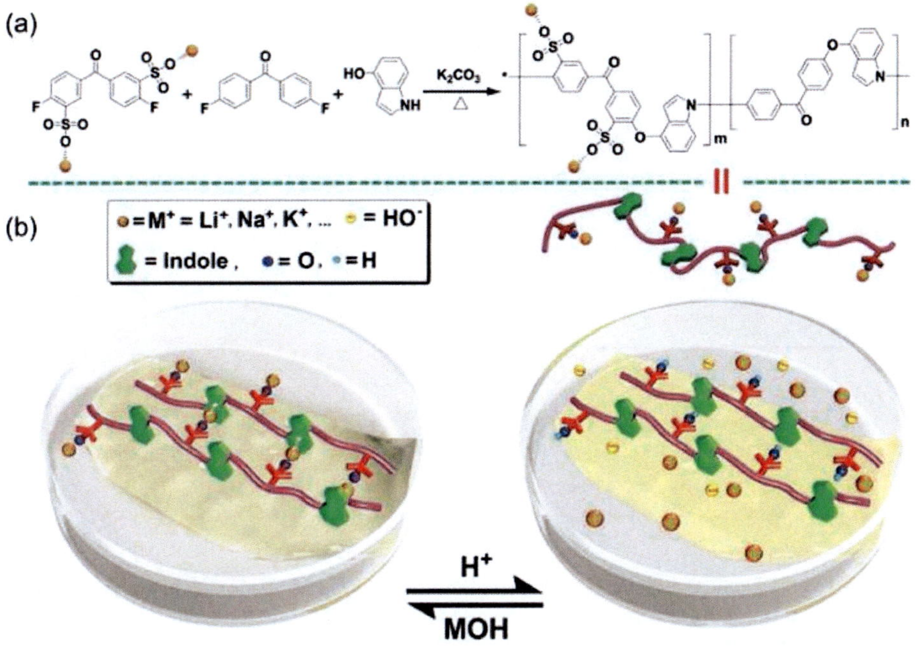

Fig. 5.11 **a** Synthetic pathway of the SPAIKs. **b** Schematic representation showing the installation and removal of cation–π complexes in response to external pH stimuli. From [10] Copyright © 2018 by John Wiley Sons, Inc. Reprinted by permission of John Wiley & Sons, Inc.

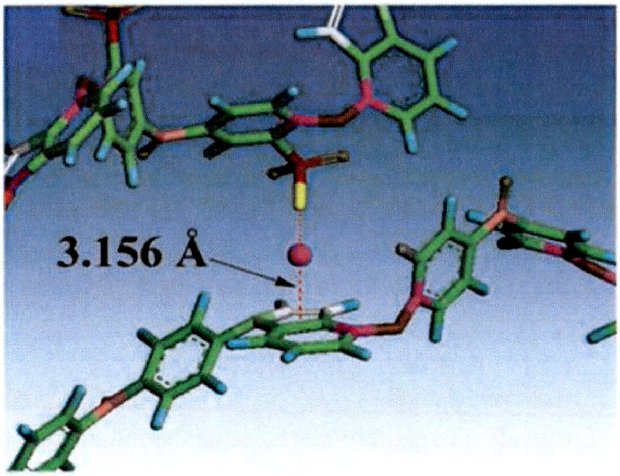

Fig. 5.12 Snapshot of two adjacent polymer chains in the simulation box under an equilibrium state. From [10] Copyright © 2018 by John Wiley Sons, Inc. Reprinted by permission of John Wiley & Sons, Inc.

indole-based difluorinated compounds (IDF) were used for copolymerization with 4,4'-dihydroxydiphenylamine (IHQ) for the preparation of sulfonated N-polyindole (SPIN) [11]. As the indole and sulfonate groups of the resulting polymer side chains exhibit efficient intermolecular cation–π interactions via metal ions, excellent thermal and mechanical properties were comparable to those of previous polymers.

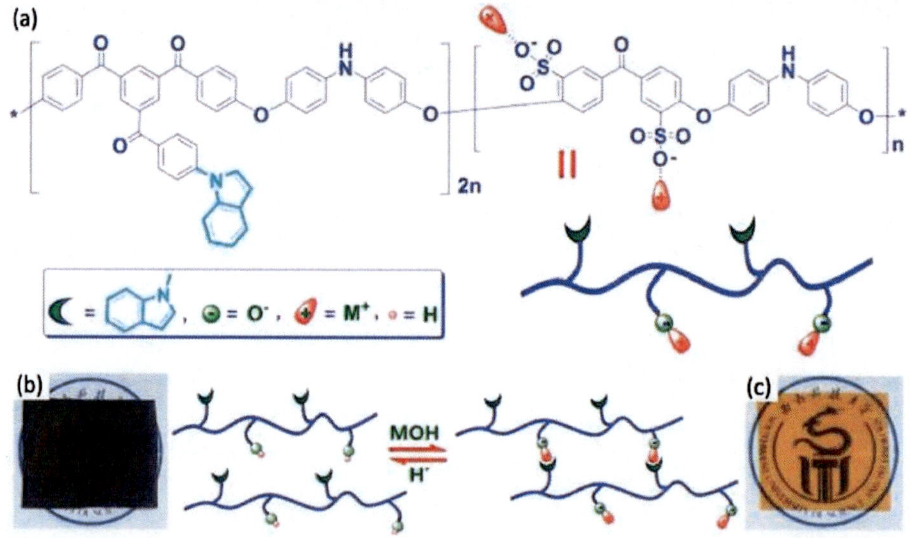

Fig. 5.13 **a** Chemical structure of SPIN. **b, c** Schematic representation showing the installation and removal of cation–π complexes in response to external pH stimuli, together with acidified and neutral SPIN films in daylight. From [11] Copyright © 2018 by John Wiley Sons, Inc. Reprinted by permission of John Wiley & Sons, Inc.

On the other hand, when the polymer film is acidified, the indole-K^+ complex is replaced by an indole-sulfonate complex, resulting in a significant decrease in thermal and mechanical properties. Interestingly, when the polymer film was immersed in a sulfuric acid solution, the film became less transparent and darkened (Fig. 5.13b and c). Furthermore, the acidified film is soluble in many organic solvents, and this supramolecular polymer can be easily recycled by reversible attachment and removal of the cation-π interactions.

5.1.2 *Surface Modification of Polymers*

5.1.2.1 Positively Charged Biointerfaces

Positively charged biointerfaces can simultaneously improve the antimicrobial and osteoinductive properties of orthopedic implants. Shi et al. developed a positively charged biointerface by the immobilization of a cationic polymer, polyhexamethylene biguanidine (PHMB), onto conjugated polydopamine (PDA) [12]. This material shows much better cytocompatibility than does free PHMB in terms of the cell half lethal dose. Both PD1 (PHMB/DA = 1:1) and PD10 (PHMB/DA = 10:1) have excellent antibacterial ability against Staphylococcus aureus (Fig. 5.14): PD1 can kill 99.9% of the bacteria at an initial concentration of 1.83×10^4 cfu/cm^2 and PD10 allows few bacteria to survive with a concentration of only 5.41×10^6 cfu/cm^2. As

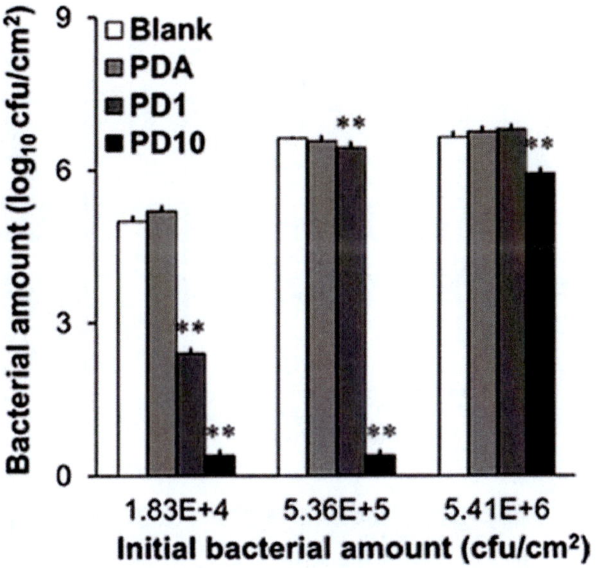

Fig. 5.14 Antibacterial properties of cationic surfaces. Inhibition of S. aureus growth at various initial concentrations on surface coatings after incubation for 24 h. Reproduced from Ref. [12] with permission from the Royal Society of Chemistry

PDA has no antibacterial effect on Staphylococcus aureus, the superior antibacterial ability of PD1 and PD10 is thought to be attributed to the surface potential resulting from the effective immobilization of PHMD by cation–π interactions (Fig. 5.15c). In other words, the anionic microbial membrane is attracted to the cationic biointerface, and the Coulombic force destroys the bacterial membrane and ultimately kills the bacteria.

The presence of cation–π interactions in PD1 and PD10 has been revealed by FTIR and NMR measurements of PHMB, PDA, PD1, and PD10. As shown in the FTIR spectra (Fig. 5.15a), the red shift of the absorption band of the aromatic ring from 1641 cm^{-1} (PD10), 1635 cm^{-1} (PD1), and 1627 cm^{-1} (PDA), and the weakened peaks at 1504 cm^{-1} in PD1 and PD10 indicate an interaction between the NH^+ of PHMB and the electron-rich aromatic ring of PDA. Solid-state ^{1}H-NMR spectra (Fig. 5.15b) showed a large upfield shift (1.169 ppm) of the aromatic ring of PD10 compared to those of PHMB (2.584 ppm) and PDA (4.742 ppm). The same shielding effect was also observed for PD1, but the intensity of PD1 was much lower than that of PD10 due to the lower PHMB content.

5.1.2.2 Adhesion Between Polymers

Direct adhesion between solid polymers usually involves chain diffusion and/or entanglement and non-covalent interactions at the interface. Wang and Xie synthesized cross-linked polystyrene (cPS), surface-sulfonated cross-linked polystyrene (ScPS) with sulfonic acid group modification on the surface of cPS [13], and the sodium salt of ScPS (SScPS) (Fig. 5.16), to clarify the involvement of cation–π interactions in the direct adhesion between polymers.

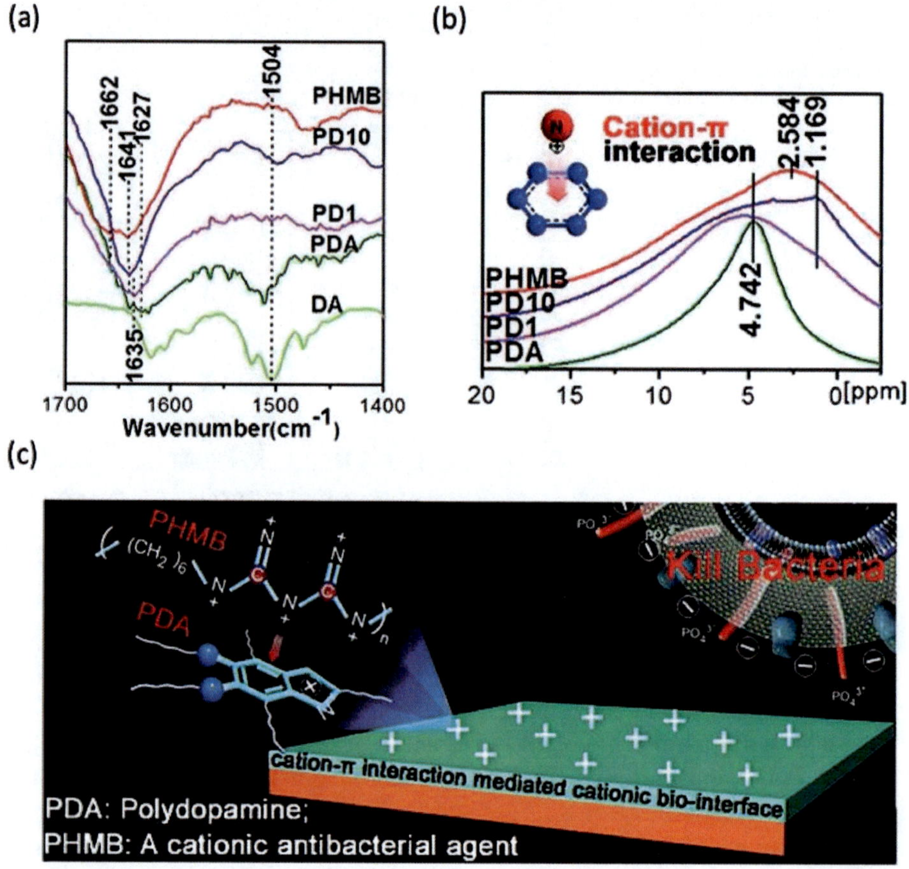

Fig. 5.15 **a** FTIR spectra. **b** Solid-state 1H-NMR and illustration of the cation–π interaction. The PHMB powders were freeze-dried from 20% PHMB solution. PDA, PD1, and PD10 powders were obtained by scratching from the corresponding samples. **c** Illustrated interaction between the cationic biointerface and bacteria. Reproduced from Ref. [12] with permission from the Royal Society of Chemistry

Epoxy polymers were selected as solid-state adhesives to study the direct bonding with the three polymeric substrates possessing the chemical functional groups described above (Fig. 5.16). The direct bonding was evaluated in tensile mode, and the adhesion forces were measured (cPS: 380×102 Ncm^{-2}, ScPS: 645+141 Ncm^{-2}, SScPS: 721+218 Ncm^{-2}). In the case of ScPS and SScPS, high adhesion values as well as cohesive failure on both epoxy and substrate surfaces were observed. XPS analysis confirmed the cohesive failure mechanism. As shown in Fig. 5.17, a carbon-only region (lower left XPS spectrum) was detected on both polymer surfaces after the adhesion test. The carbon-only regions correspond to the non-sulfonated bulk (cPS) inside the SScPS, suggesting the occurrence of bulk fracture in the SScPS. In addition, sulfur and sodium were no longer detected in the fracture region of the

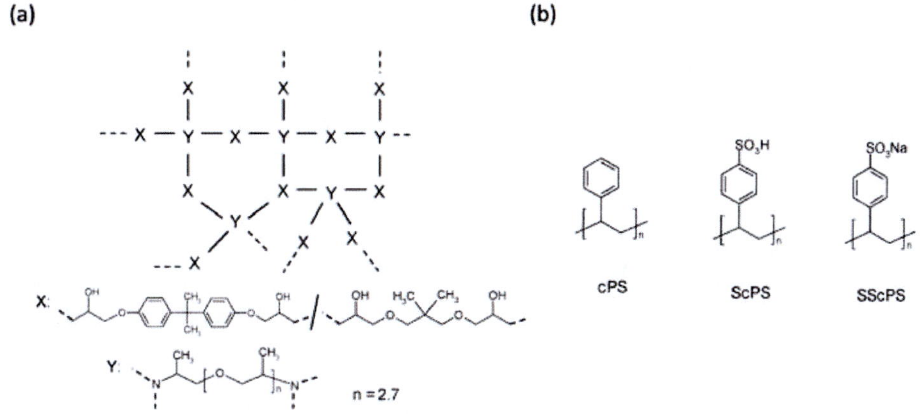

Fig. 5.16 Chemical structures of **a** thermoset epoxy polymer and **b** substrate polymers. Reproduced from Ref. [13] with permission from the Royal Society of Chemistry

SScPS sample, with nitrogen detected instead. This suggests the transfer of nitrogen-containing epoxy polymer to the SScPS surface and the cohesive breakdown of the epoxy polymer.

As the epoxy polymers are highly cross-linked, no chain diffusion occurs at the interface, and no Coulomb interaction or strong hydrogen bonding interaction occurs between the epoxy polymer and SScPS due to the absence of acidic protons. The only major contributing force responsible for the macroscopic high adhesion are the cation–π interactions between the abundant sodium ions of SScPS and the aromatic rings of the epoxy surface. The cohesive breakdown of the adhesive bonds of SScPS means that the collective strength of the multiple cation–π interactions is extremely large. Such strong and direct bonding is of real importance, as adhesion can be obtained within minutes without the need for the curing required by conventional adhesives.

5.1.2.3 Copolymer Synthesis on Negatively Charged Surfaces

Fan et al. found a new method for the preparation of copolymers with adjacent cation–aromatic sequences by free radical polymerization using a cation–π complex [14]. Free radical polymerization of two pairs with the same vinyl head, a quaternary N-type cationic monomer (2-(acryloyloxy)ethyltrimethylammonium chloride (ATAC)) and a phenyl-terminated aromatic monomer (2-phenoxyethylacrylate (PEA)) in equimolar ratios, was carried out to study the conversion of the copolymers (Fig. 5.18). The cationic and aromatic monomers showed marked differences in conversion rates at low concentrations (<1.0 M), but the difference was reduced with increases in the total concentration. At a high concentration (1.0 M), the conversion rates of the cationic and aromatic monomers were the same throughout the reaction,

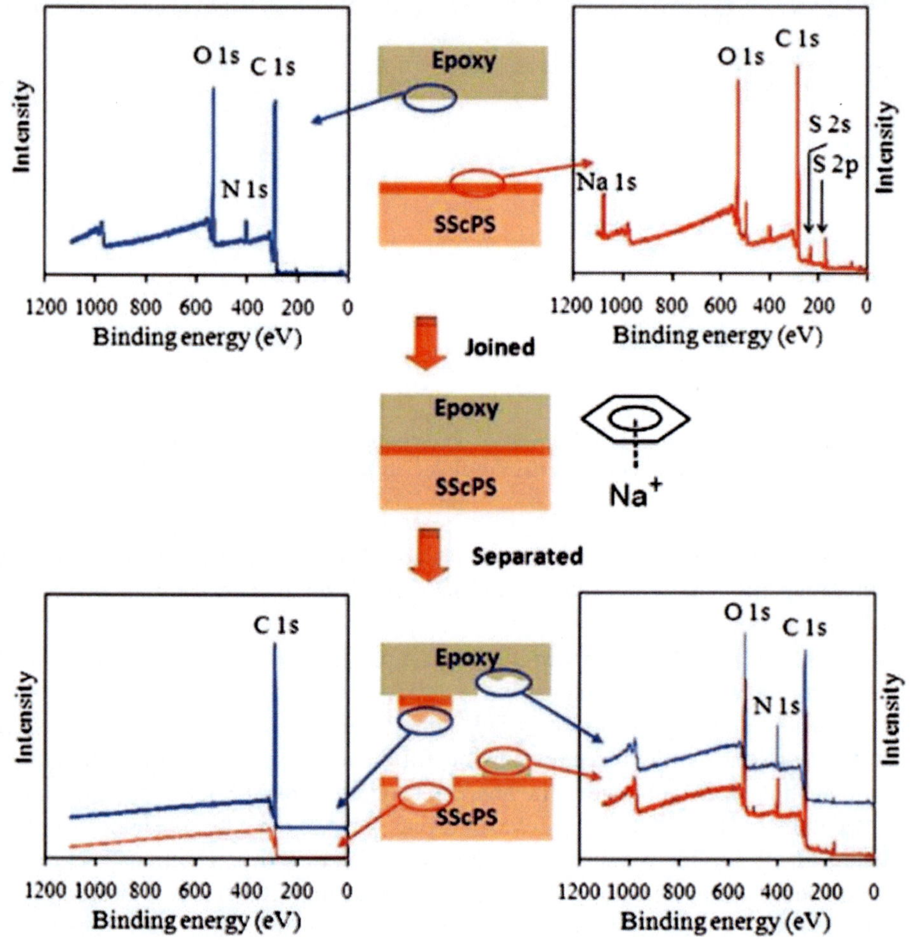

Fig. 5.17 XPS surface analyses of SScPS and the epoxy polymer before adhesion bonding and after the adhesion test. Reproduced from Ref. [13] with permission from the Royal Society of Chemistry

indicating that the two monomers were equally incorporated into the copolymer (Fig. 5.19).

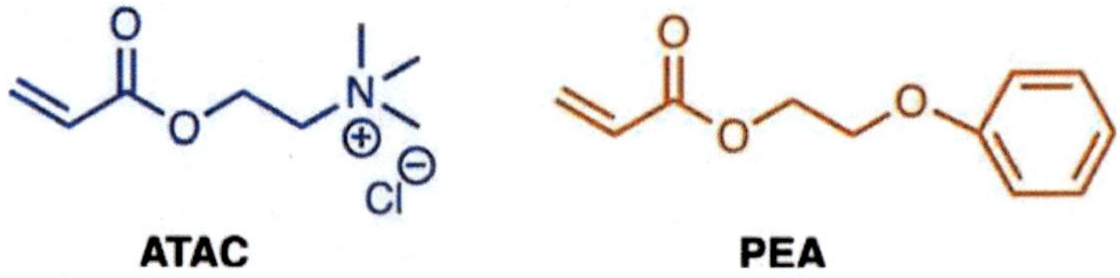

Fig. 5.18 Chemical structures of two typical monomers used in this study. Cationic monomer, 2-(acryloyloxy)ethyl trimethylammonium chloride (ATAC), and aromatic monomer, 2-phenoxyethyl acrylate (PEA)

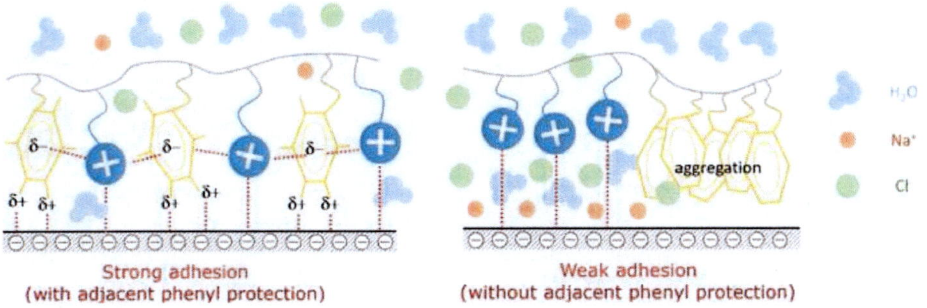

Fig. 5.19 Schematic illustration of the design strategy. Cation–π complex-aided free-radical polymerization for the synthesis of poly with adjacent cationic–aromatic sequences and its supramolecular hydrogel showing electrostatic adhesion in seawater. The counter ions and salt ions are not included for simplicity. Reprinted by permission from Springer Nature [14], Copyright Springer Nature Singapore Pte Ltd. 2019

This sequence-controlled copolymer composed of cationic/aromatic monomers is capable of fast, strong, and reversible adhesion to negatively charged surfaces in salt water, such as in physiological and marine environments. The aromatic rings on the copolymers were also found to enhance the electrostatic interaction of adjacent cationic residues to the surfaces, even in highly ionic media. These findings should contribute to the development of adhesives in saline solution.

5.1.2.4 Wettability Switch

The control of surface wettability by ionic liquids (ILs) is important for the design of intelligent materials and devices. Chang et al. prepared a mixed molecular brush composed of polyphenylethyl methacrylate and 1H, 1H, 2H, 2H-perfluorodecyltrimethoxysilane (PPhEtMA-co-PFDMS) (Fig. 5.20a) and found that this wettability could be switched by IL, (1-ethyl-3-methylimidazolium bis(trifluoromethylsulfonyl)imide ([EMIm] [NTf_2]) [15]. This change in wettability was attributed to the temperature-dominated cation–π interaction between [EMIm][NTf_2] and PPhEtMA-co-PFDMS brushes.

At low temperatures, [EMIm] [NTf_2] forms a specific solvation structure with the PPhEtMA chain (Fig. 5.20b). The [EMIm] cation interacts preferentially with the benzene face of the PPhEtMA chain through cation–π interactions. In this case, the PPhEtMA chain is in a solvated state and extends outward to shield the PFDMS and form a lipophilic surface.

At 20 °C, the [EMIm] [NTf_2] contact angle (CA) of a silicon micropost (SiMP) modified with PPhEtMA-co-PFDMS is 10.0±0.5° (Fig. 5.20b). On the other hand, when the temperature is raised to 100 °C, the PPhEtMA chain is desolvated and [EMIm] [NTf_2] is released from the PPhEtMA chain, thereby exposing the PFDMS and leading to its hydrophobicity. The CA in this case is 127.8±1.8° (Fig. 5.20b). Therefore, by controlling the cation–π interactions between [EMIm]

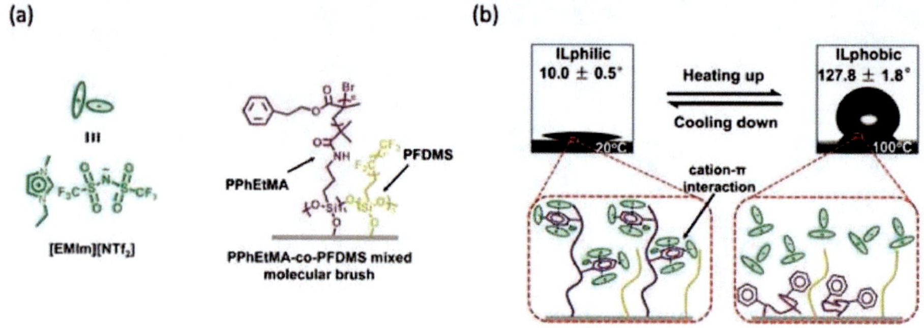

Fig. 5.20 **a** Chemical structures of [EMIm][NTf_2] and the PPhEtMA-co-PFDMS mixed molecular brushes on the surface. **b** At 20 °C, [EMIm][NTf_2] forms a specific solvation structure with the side benzene group on the PPhEtMA chains through cation–π interactions, resulting in an ILphilic surface with an [EMIm][NTf_2] contact angle (CA) of about 10.0°. When the temperature is increased to 100 °C, the PPhEtMA chains collapse and [EMIm][NTf_2] is liberated from the polymer chains, exposing the PFDMS moiety and leading to an ILphobic surface with an [EMIm][NTf_2] CA of about 127.8°. Reproduced from Ref. [15] with permission from the Royal Society of Chemistry

[NTf_2] and the PPhEtMA-co-PFDMS brushes, reversible temperature-induced changes in wettability can be achieved.

5.1.3 *Indole-Based Hydrogels*

Chang et al. have prepared a new type of recyclable indole-based hydrogel. This hydrogel shows effective hydroxide extraction behavior due to the synergistic effect of cation–π interactions and hydrogen bonding (see, Sect. 2.5.2) [16]. More interestingly, the color change resulting from the host–guest interaction can be used to visually detect absorption performance. The 4-hydroxyindole hydrogel (4-HIG) can be conveniently synthesized from 4-hydroxyindole and formaldehyde as shown in Fig. 5.21a.

The ability of the gel to take up sodium hydroxide is indicated by the color change of phenolphthalein. The concentration of NaOH in the aqueous solution (25 mM at $t = 0$ h) was dramatically decreased by the addition of 4-HIG gel (Fig. 5.21b); after 0.5 h, the amount of NaOH decreased to about 9.0 mM, which may be due to the deprotonation of the phenolic hydroxy groups by OH anions. Even when the absorption time was extended to 24 h, the concentration of NaOH in the aqueous solution remained almost unchanged, with 91.5% of the NaOH incorporated in the gel.

The possibility of a cation–π interaction between Na^+ and the 4-hydroxyindole (4-HI) moiety was verified by UV–vis absorption spectra. The absorption spectra of 4-HI in the absence and presence of two equivalents of Na^+ and the difference spectrum between 4-HI-Na^+ and 4-HI are shown in Fig. 5.22a. The difference spectrum clearly shows a negative band at 216 nm and a positive band at 226 nm. This negative/positive

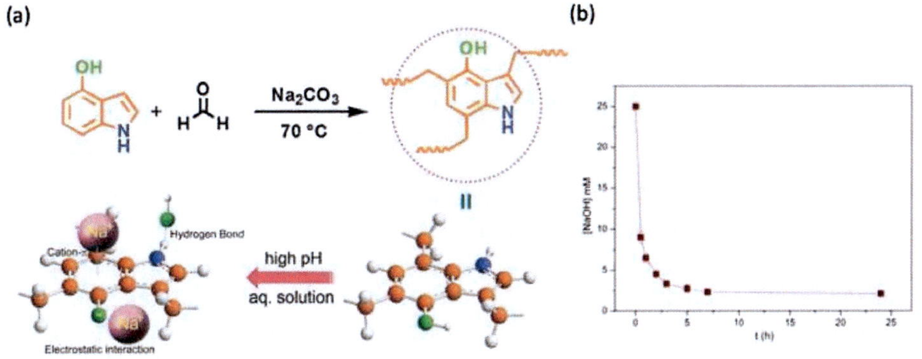

Fig. 5.21 **a** Fabrication of the 4-HIG hydrogel. Proposed absorption mechanism of the 4-HIG hydrogel in a high pH aqueous solution (color code: C, orange; H, white; N, blue; O, green). **b** The concentration of NaOH decreases as a function of time in the presence of the 4-HIG gel. These concentrations were determined by measuring the absorbance intensity of phph at 554 nm as a function of time. Reproduced from Ref. [16] with permission from the Royal Society of Chemistry

band pair can be attributed to the weakening of the strong Bb transition of the 4-HI indole ring by the cation–π interaction between Na^+ and 4-HI. The high NaOH extraction behavior is due to the synergistic effect of the cation–π interactions and hydrogen bonding (Fig. 5.22b) as mentioned in Chap. 2.

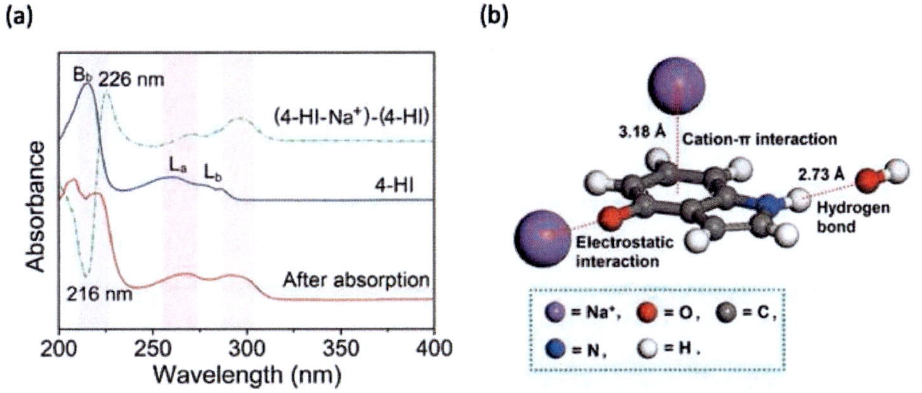

Fig. 5.22 **a** UV absorption spectra of 4-HI (50 mM) and its 1:2 complex with Na^+ at pH 7.0 (1 mM HEPES buffer). The difference spectrum for 4-HI–Na^+ minus 4-HI is doubly magnified for clarity. **b** The optimized geometry of 4-HI–Na^+ at the M062x/6–311+G** level. Reproduced from Ref. [16] with permission from the Royal Society of Chemistry

5.2 Biopolymer and Biomimetic Polymers

5.2.1 Mussel Underwater Adhesion

Mussels effectively achieve underwater adhesion to wet substrates by secreting a proteinaceous, thread-like substance called byssus. The protein involved in this adhesion is called mussel foot protein (mfp), and it has been widely studied as an important model for understanding biological adhesion in water and for developing biomimetic wet adhesives. The catechol portion of the protein, 3,4-dihydroxyphenyl-L-alanine (DOPA), is known to actively participate in the underwater adhesion mechanism of mussels (Fig. 5.23) [17]. Use of a surface force apparatus (SFA) (Fig. 5.24) to investigate the mechanism of mussel adhesion underwater elucidated that cation–π interactions, one of the strongest intermolecular interactions in water, is important to these adhesion proteins.

Lu et al. investigated the direct measurement of poly-L-lysine (PLL)—aromatic polymer adhesion energies in aqueous solutions by SFA [19]. One mica surface was coated with PLL containing an alkyl-NH_3^+ moiety, and another mica surface was coated with three polymers with aromatic side chains, polystyrene (PS), poly-L-tyrosine (PTyr), or poly-L-tryptophan (PTrp) (Fig. 5.24). The adhesion strength between PLL and one of the three aromatic polymers (PTrp, PTyr, or PS) followed the order: PLL/PTyr <PLL/PS <PLL/PTrp, but did not match the quadrupole moment of the probed π-system. When Li^+, Na^+, K^+, NH_4^+, and -NR_3^+ were then injected between the PTrp and PLL-coated surfaces, the cation–indole interaction energies were in the order: Li^+, Na^+ <K^+, NH_4^+ < -NR_3^+.

In general, the strength of the cation–π interactions depends on the quadrupole moment of the π system. However, in this experiment, PS (quadrupole moment = −8.2) showed a higher affinity for the positively charged PLL than for PTyr (quadrupole moment = −8.5). These results suggest that, in aqueous solution,

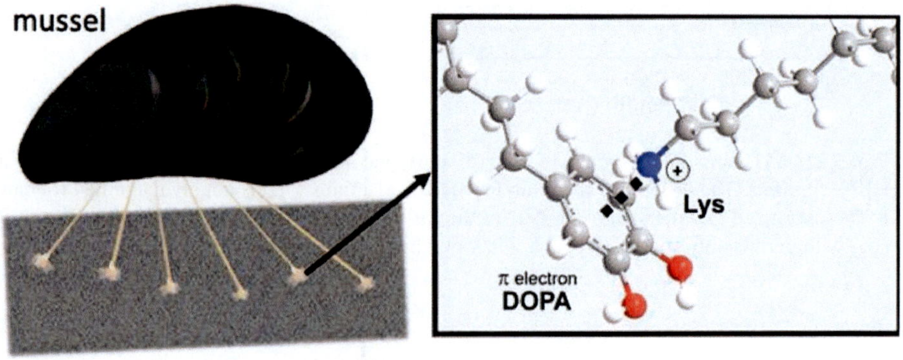

Fig. 5.23 Picture of a mussel with foot proteins containing DOPA. Reprinted with permission from [17]. Copyright 2019 American Chemical Society

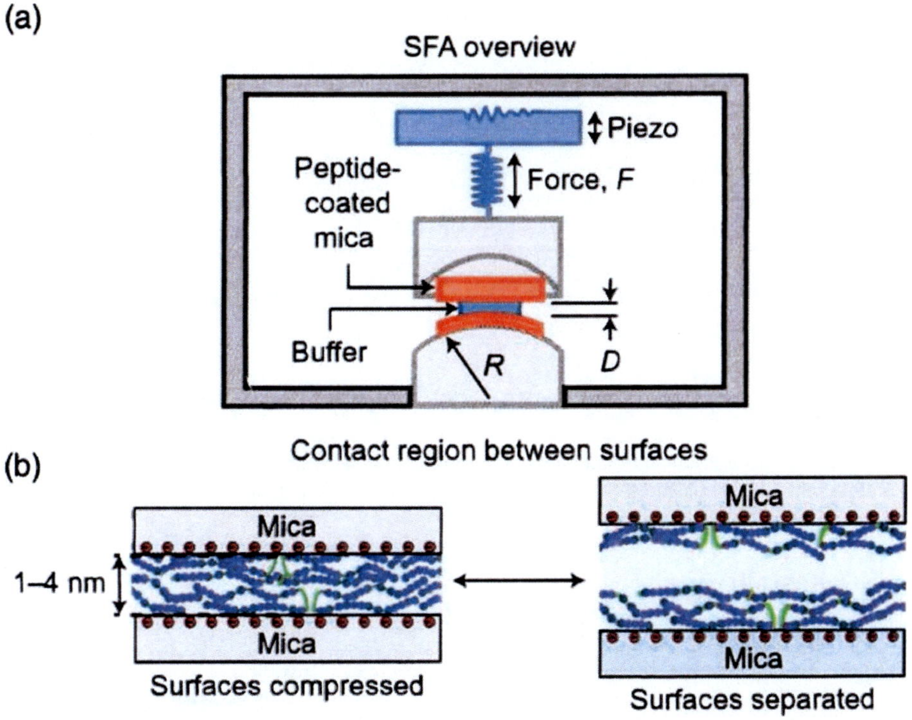

Fig. 5.24 **a** Schematic of the SFA experimental setup. The peptide was solution deposited onto mica surfaces before placing it into the SFA. **b** Compression of a multilayer peptide film and measurement of the interaction energies during the separation of the surfaces. Reprinted by permission from Springer Nature [18], Copyright Springer Nature Singapore Pte Ltd. 2017

the composition and strength of the cationic hydration shell have a greater influence on the cation–π interactions than does the cationic charge density (Fig. 5.25).

Gebbie et al. studied the aggregation of model peptides containing Lys and aromatic side chains (Phe, Tyr, or DOPA) via cation–π interactions and found that the cohesion followed the order: DOPA<Tyr<Phe [18]. These results are not consistent with the order of the magnitude of the quadrupole moment similar to the above results, which suggests that the presence of hydroxy groups on the aromatic ring decreases the number of anions that can be bound, thereby decreasing the configuration entropy. As a result, the cohesion of Phe-containing peptides is higher than that of DOPA or Tyr-containing peptides.

In the recombinant Rmfp-1 (Fig. 5.26a), all DOPA residues in the foot protein were replaced by Tyr, and the contribution of the cation–π interactions was measured by SFA. The interaction force (Wco) between two Rmfp-1 layers on mica was measured as ~2.9 mJ/m^2 in 0.1 M acetic acid (pH 3.0), which was close to the interaction force between pvfp-1 (green mussel) or mcfp-1 (blue mussel) layers (Wco ≈ 3.5 mJ/m^2 and Wco ≈ 2.5 mJ/m^2, respectively).

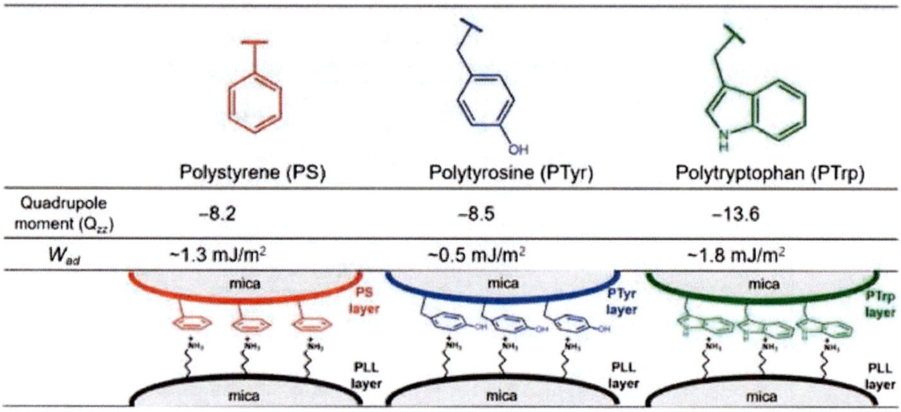

Fig. 5.25 PLL-aromatic polymer (PS, PTyr, or PTrp) adhesion energies in aqueous solution. Reprinted with permission from [17]. Copyright 2019 American Chemical Society

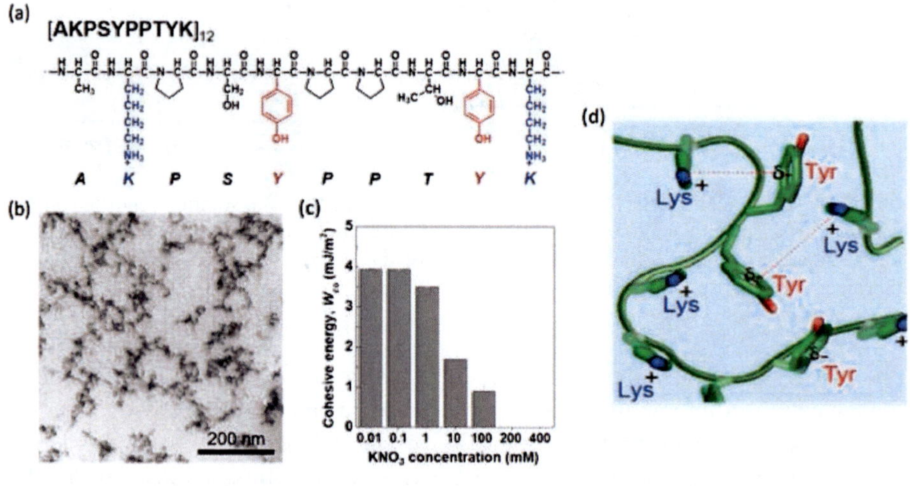

Fig. 5.26 **a** Primary sequence of Rmfp-1 and **b** aggregation of this protein in an aqueous solution observed by cryo-TEM. **c** Effect of K^+ concentration on the cohesive energy of Rmfp-1. **d** Contribution of cation–π interactions in the aggregation of the protein. Reprinted with permission from [20]. Copyright 2017 American Chemical Society

When the concentration of K^+, a cation that competes with Lys, was increased from 0 to 400 mM, the cohesive force in the Rmfp-1 membrane decayed exponentially, eventually reaching a value of 0 mJ/m^2 (Fig. 5.26c). This result indicates that the cation–π interactions between the Tyr residues and the positively charged Lys residues contribute to the strong cohesion (Fig. 5.26d) [19].

To confirm the aggregation of Rmfp-1 in an aqueous medium, a deionized aqueous solution of the protein (1 mg/mL) was observed by cryo-TEM and Rmfp-1 was found to aggregate in aqueous solution similar to the case of mfp-1 from M. edulis (Fig. 5.26b). As this is a cationic polyelectrolyte with an equal content of Tyr and Lys (20 mol%) without negatively charged residues in its primary sequence, it was concluded that cation–π interactions between the Tyr and Lys residues are responsible for the above aggregation (Fig. 5.26d) [20].

5.2.2 *Mussel-Inspired Polymers*

Lim et al. used two opposing mica surfaces coated with polycatecholamine, produced by the oxidation of norepinephrine, to evaluate the synergistic effects of catechol and amines. SFA was used to measure the cohesion between the two layers, with the pyrocatechol-coated layer as the control substance [21]. As a control material, a layer of pyrocatechol coating was prepared. Surprisingly, the cohesion between the poly(catecholamine) coatings (Wco (cohesion)≈ 6.4 mJ/m^2)(interaction energy per unit area, W, between two flat surfaces, defined by W=F/1.5πR) was about 30 times higher than that between the two poly(catechol) coatings (Wco ≈ 0.2 mJ/m^2) (Fig. 5.27). This may be due to the hydrogen bonding between the acidic phenolic hydroxy groups and the amino groups, and the cation–π interactions between the ionized ammonium and the aromatic rings.

Poly(norepinephrine) has attracted much attention as a versatile coating agent for the control of surface properties in the field of membrane systems such as water treatment, drug delivery, biosensors, energy storage, tissue engineering, and adhesives.

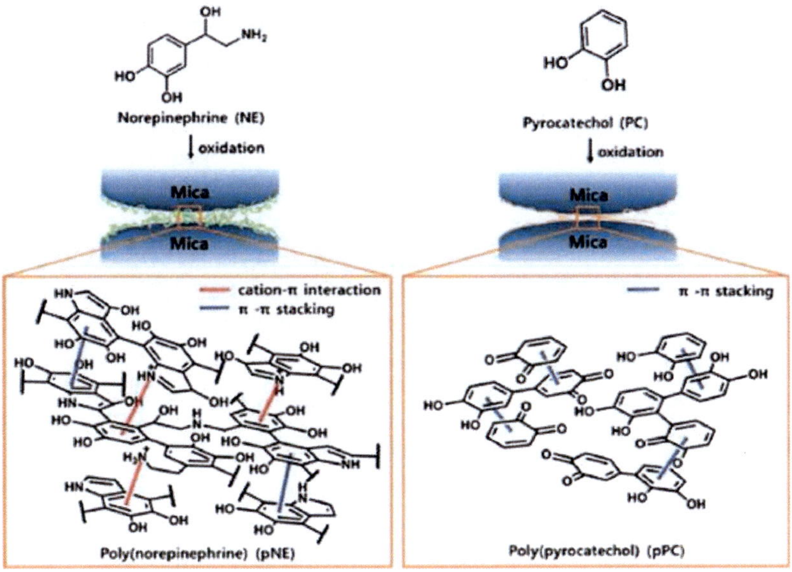

Fig. 5.27 Chemical structure and schematic representation of surface force measurements of pNE and pPC by SFA. From [21] Copyright © 2016 by John Wiley Sons, Inc. Reprinted by permission of John Wiley & Sons, Inc.

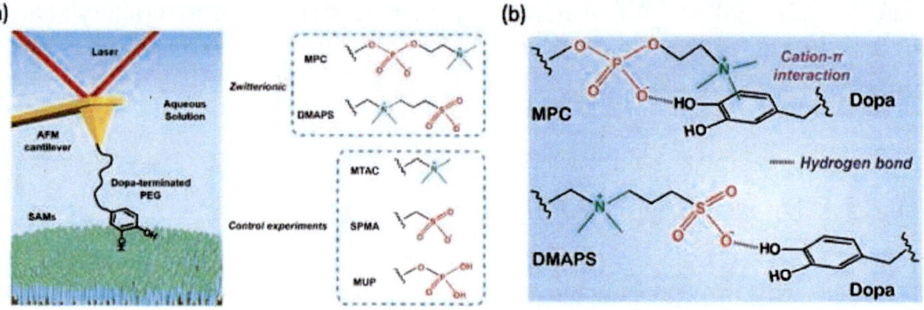

Fig. 5.28 **a** Schematic illustration of the AFM experimental setup for measuring the interaction between a single Dopa-terminated molecule and SAMs. **b** Schematic illustration of the binding mechanism for the zwitterion–Dopa interaction. Reproduced from Ref. [22] with permission from the Royal Society of Chemistry

Xie et al. designed methacrylic acid derivatives with MPC and DMAPS groups based on the adhesion mechanism of the purple mussel described above, and investigated their interaction with Dopa [22]. MPC and DMAPS shown in Fig. 5.28b and PEG polymers terminated with Dopa were used as AFM tips to measure the interaction energy. Figure 5.28a shows a schematic representation of the experimental setup for measuring the interaction between SAMs and a single Dopa-terminated molecule immobilized on an AFM chip.

Based on the measurements obtained, the calculated bond dissociation energies are summarized in Table 5.1. Interestingly, the dissociation energy of a single Dopa molecule interacting with MPC is calculated to be ~19.4 kBT, which is almost twice the dissociation energy of DMAPS (~10.8 kBT). The dissociation energy of Dopa interacting with MTAC is calculated to be ~11.8 kBT and is attributed to the cation–π interaction between the -$N^+(CH_3)_3$ group of MTAC and the catechol π-ring of Dopa. On the other hand, the dissociation energy of the hydrogen bond formed between the phosphate group of MUP and the catechol hydroxy group of Dopa is calculated to be ~11.3 kBT. As expected, the sum of the dissociation energies for the cation–π interaction (~11.8 kBT) and the hydrogen bond (~11.3 kBT) is very close to that of the MPC-Dopa interaction (~19.4 kBT). Therefore, the MPC-Dopa interaction is most likely due to hydrogen bonding and synergistic cation–π interactions between the -$N^+(CH_3)_3$ group and the catechol moiety.

Table 5.1 Calculated bond dissociation energies based on AFM-SMFS results at pH 4

Molecules interacting with Dopa	Surface ionic group	Free-energy barrier (kBT)
MPC	-O_3-$PO(CH_2)N^+(CH_3)_3$	19.4
DMAPS	-$(CH_3)_2$ $N^+(CH_2)_3SO_3$-	10.8
MTAC	-$N^+(CH_3)_3$	11.8
SPMA	-SO_3^-	10.9
MUP	-$OPO(OH)_3$	11.3

On the basis of these findings, poly(MPC-co-DMA), which has both MPC and Dopa moieties in the side chain, was developed as a promising molecular framework for underwater adhesives. The interaction forces between the poly(MPC-co-DMA) solution and the two mica surfaces were measured at the nanoscale level by SFA, and the adhesion energy was found to gradually increase to reach 14.1 mJ/m^2 after 60 min, which is much larger than the adhesion energy of Rmfp-1 with decapeptide (3–6 mJ/m^2).

5.2.3 Lignin Biopolymers

The adsorption properties of alkaline lignin on gold surfaces coated with cationic polymers have been investigated using quartz crystal microbalance in aqueous and organic solvents (Fig. 5.29) [23]. The adsorption of PDDA (poly(diallyldimethylammonium chloride)) onto a gold-coated quartz crystal causes an immediate decrease in frequency. This frequency change is related to the effective mass of the PDDA adsorbed on the surface of the gold crystal and the water closely associated with the PDDA. PDDA adsorption shows a large change at a frequency of −4.9 Hz at pH 7. An alkaline lignin solution was adsorbed on the PDDA surface at increasing concentrations; the final change in frequency of lignin film on PDDA was −21 Hz, and the corresponding mass of the alkaline lignin was 3.6 mg/m^2.

A comparison of the FTIR spectra of a dried mixture of alkali lignin and PDDA with the spectra of the individual components reveals that the main aromatic peak of the alkali lignin at 1513.7 cm^{-1} shifts to 1515.6 cm^{-1} when the lignin is complexed with PDDA (Fig. 5.30). This change in peak position is attributed to the adsorption of the quaternary ammonium surfactant on the aromatic surface of the polyphenols. Substitution of the phenolic hydroxy group with a hydrophobic methyl group does not alter the adsorption property of lignin on the PDDA surface, indicating that hydrogen bonding is not involved in the adsorption. Therefore, these results strongly suggest the existence of a cation–π interaction between the aromatic group of the alkaline lignin and the cationic moiety of PDDA.

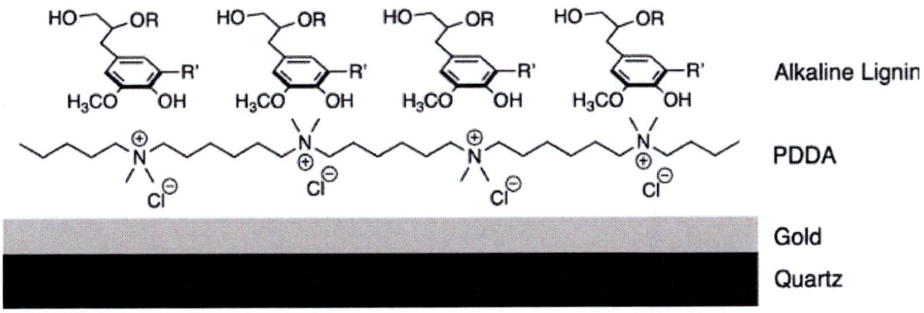

Fig. 5.29 Schematic representation of interactions between alkaline lignin and PDDA on a gold surface

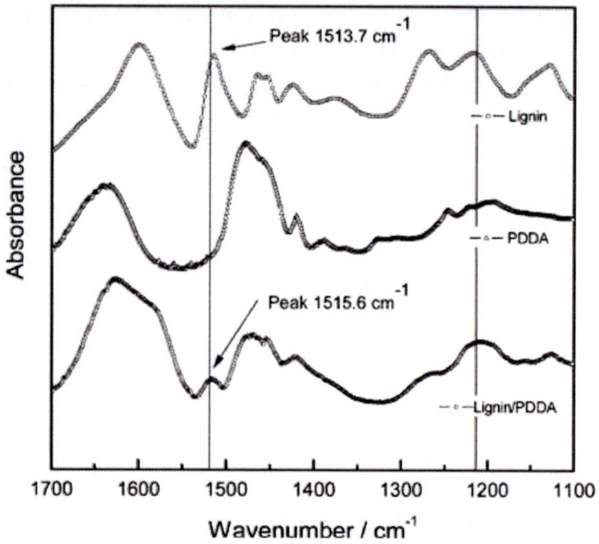

Fig. 5.30 FTIR transmission spectra of alkali lignin, PDDA, and a dried alkali lignin-PDDA complex. Note the shift related to the aromatic skeletal vibration at 1515 cm^{-1}. Reprinted with permission from [23]. Copyright 2009 American Chemical Society

Sodium lignosulfonate (SL) is generally present as a byproduct in spent sulfite pulping liquor from the paper and pulping industry. By using the SL fraction as a polyanion and poly(diallyldimethylammonium chloride) (PDAC=PDDA) as a polycation, layer-by-layer self-assembled multilayer films were constructed. There is minimal adsorption of SL in a solution on the PDAC surface without NaCl, while SL (Mw = 18700 Da) in the solution with NaCl can form SL/PDAC multilayer films [24].

Fluorescence spectroscopy was carried out to clarify the cation–π interaction between SL and PDAC in the presence of sodium ions. Figure 5.31 shows the fluorescence emission spectra of SL, PDAC, and the SL/PDAC, SL/NaCl, and SL/PDAC/NaCl mixtures. The maximum emission wavelength for both the SL and PDAC solutions is 530 nm. When PDAC is added to the SL solution, the spectrum peak of SL shifts to 530–500 nm. This indicates the presence of a cation–π interaction between PDAC and the aromatic groups of SL. On the other hand, when NaCl is added to the SL/PDAC solution, the intensity of the emission at 500 nm increases, indicating that the addition of NaCl enhances the cation–π interaction between PDAC and SL.

5.3 Graphene-Based Materials

Graphene is an aromatic compound with a two-dimensional spread of benzene rings that can be regarded as a new aromatic polymer. Its unique structure confers a variety of superior physicochemical, electronic, optical, thermal, and mechanical properties, leading to a wide range of potential applications.

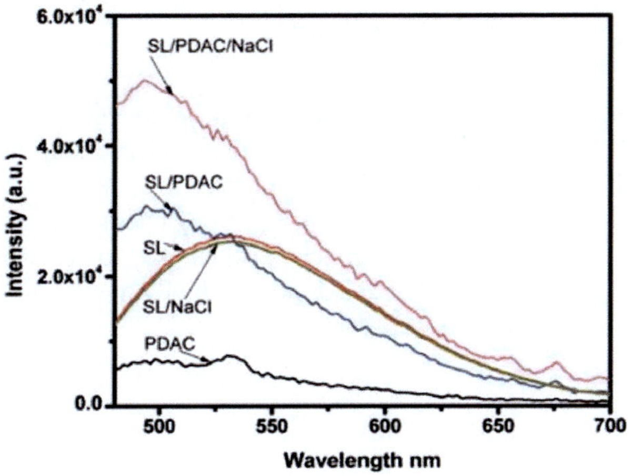

Fig. 5.31 Fluorescent emission spectra of SL, PDAC, SL/PDAC mixture, SL/NaCl mixture, and SL/PDAC/NaCl mixture: SL, 0.8 g/L; PDAC, 0.4 mmol/L; NaCl, 0.1 mol/L. Reprinted with permission from [24]. Copyright 2011 American Chemical Society

Graphene oxide (GO) is obtained by oxidation of graphene, in which various oxygen-containing functional groups, such as hydroxy, epoxide, carboxy, and carbonyl groups, are covalently attached to the aromatic ring (Fig. 5.32). The cross-linking with cations by cation–π interactions contributes to the stability of the membrane structure, the control of the interlayer spacing, and the membrane selectivity. Furthermore, graphene has been used in the separation of cations, adsorption of organic compounds, preparation of novel materials, energy generation and storage, and so on.

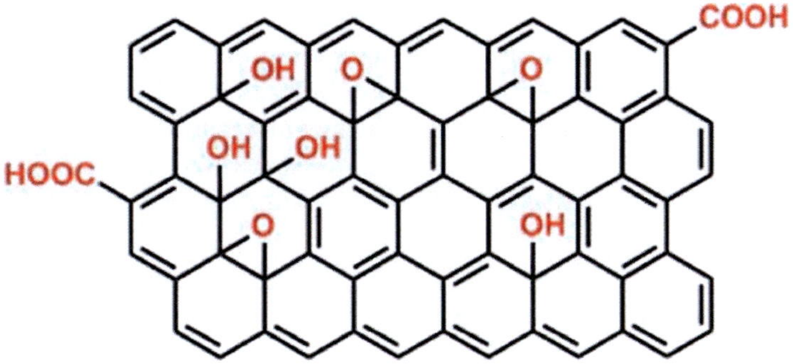

Fig. 5.32 Structure of graphene oxide showing possible oxygen-containing functionalities. From [25] Copyright © 2020 by John Wiley Sons, Inc. Reprinted by permission of John Wiley & Sons, Inc.

5.3.1 *Graphene Oxide Membranes*

Cation–π interactions contribute to the adsorption of metal ions and charged contaminants onto GO sheets. When metal cations are introduced into the middle layer of stacked GO sheets, the uniformity of the GO films in aqueous solution is increased by the cross-linking of GO sheets through the strong interaction between the cations and the GO sheets. This cross-linking can also be used to allow precise control of the interlayer spacing of lamellar GO films. Such GO membranes can be used for the separation and adsorption of organic cations.

5.3.1.1 Separation of Metal Ions

It has been reported that the alkali and alkaline earth cation transport properties of GO films vary with the type of ion [26]. For nanosized GO membranes, at 20 °C, the relative transport rates of cations were found to be: $H^+ > Mg^{2+} > Na^+ > Ba^{2+}$, Ca^{2+} and K^+. The interaction between the hydrated K^+ cluster and the π cluster is strong enough to result in partial desolvation of the K^+ ions, indicating that the transport rate of K^+ is much smaller than that of Na^+ (Fig. 5.33). Similarly, for alkaline earth cations (Mg^{2+}, Ca^{2+}, and Ba^{2+}), desolvation is influenced by the balance between the electrostatic interaction between the cation and surrounding water molecules and the cation–π interaction between the metal ion and the sp^2 cluster. In the case of Mg^{2+}, as the ion size is much smaller than that of Ca^{2+} or Ba^{2+}, the hydration energy of Mg^{2+} ion is much larger. Therefore, a sufficient amount of water molecules attaches to the Mg^{2+} ions in the GO nanocapillary, thus preventing the sp^2 (graphene) clusters from interacting with the Mg^{2+} ions.

Sun et al. prepared three types of graphene oxide (GO) with hydroxy, amino, and carboxy groups on surface and investigated the permeability of Cu^{2+} and Mg^{2+} (Fig. 5.34) [27]. In the case of carboxy GO, a significant difference was observed between them: Mg^{2+} showed better permeability than did Cu^{2+}, while Cu^{2+} was found to be blocked. This may be due to the fact that transition metal cations tend to interact with the C-O moiety via coordination interactions, while alkali and alkaline earth cations interact with the sp^2 zone via cation–π interactions.

5.3.1.2 Interlayer Spacing Control

Cation–π interactions can be used to control the interlayer spacing of GO films with angstrom-level accuracy. As shown in Fig. 5.35b, the GO membranes prepared from various salt solutions have different surface spacing values. When they are immersed in KCl solution, the spacing becomes K^+-specific in all cases, indicating that other ions can be effectively excluded (Fig. 5.35c) [28]. This clearly shows that the facial spacing of the GO film is controlled by the strong cation–π interactions between the potassium ion and the upper and lower GO layers (Fig. 5.35a).

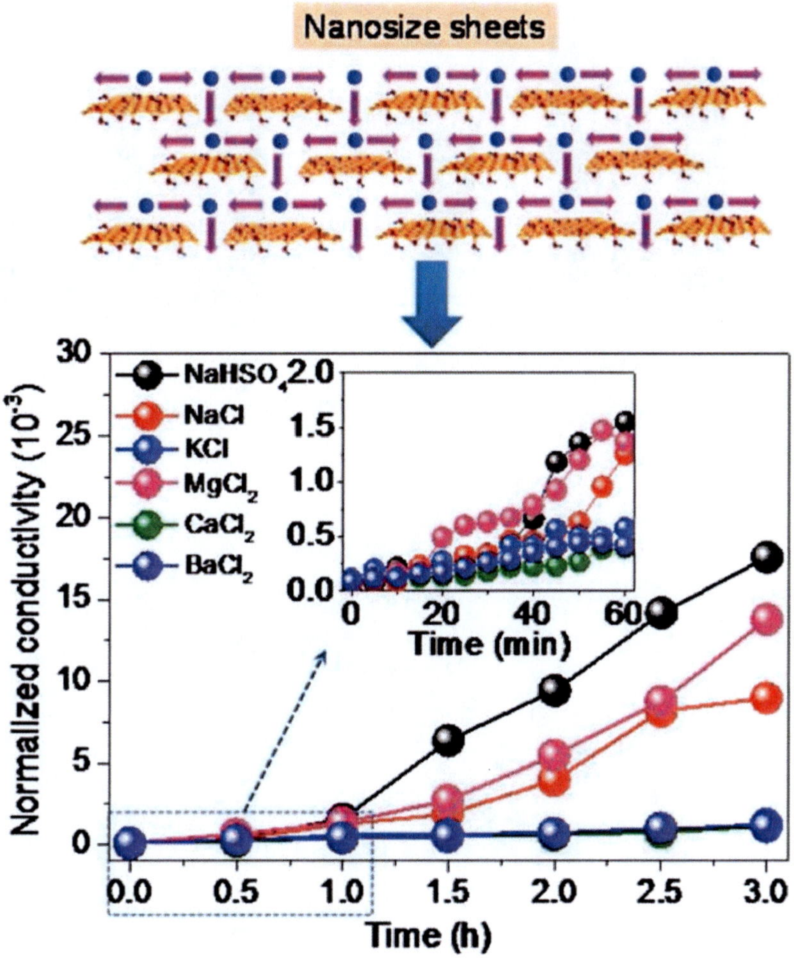

Fig. 5.33 Schematics of the lamellar structures (top panels) and normalized conductivities of the filtrates of studied cations (bottom panels) for GO membranes composed of nanosized GO sheets. Reprinted with permission from [26]. Copyright 2014 American Chemical Society

5.3.1.3 Separation of Organic Solvents

Wang et al. prepared selective permeable GO-based membranes for a variety of solvents through utilization of the interactions between metal cations and GO [29]. Use of inorganic salts can modify the permeation properties of the GO membranes for various solvents. Figure 5.36 shows only minimal permeation of saturated aliphatic alcohols through the modified GO membrane, whereas solvents with lone pair electrons, such as acetone, acetonitrile, and dichloromethane, are able to pass through the membrane quickly. This may be due to electrostatic forces and hydrogen bonding between the solvent and the metal ions as shown in Fig. 5.37.

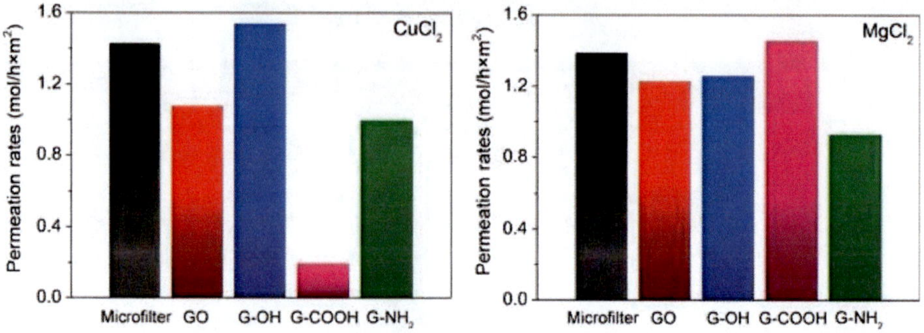

Fig. 5.34 Cu^{2+} and Mg^{2+} permeation rates through blank microfilters, GO, G—OH, G—COOH, and G—NH_2 membranes. Reprinted with permission from [27]. Copyright 2014 American Chemical Society

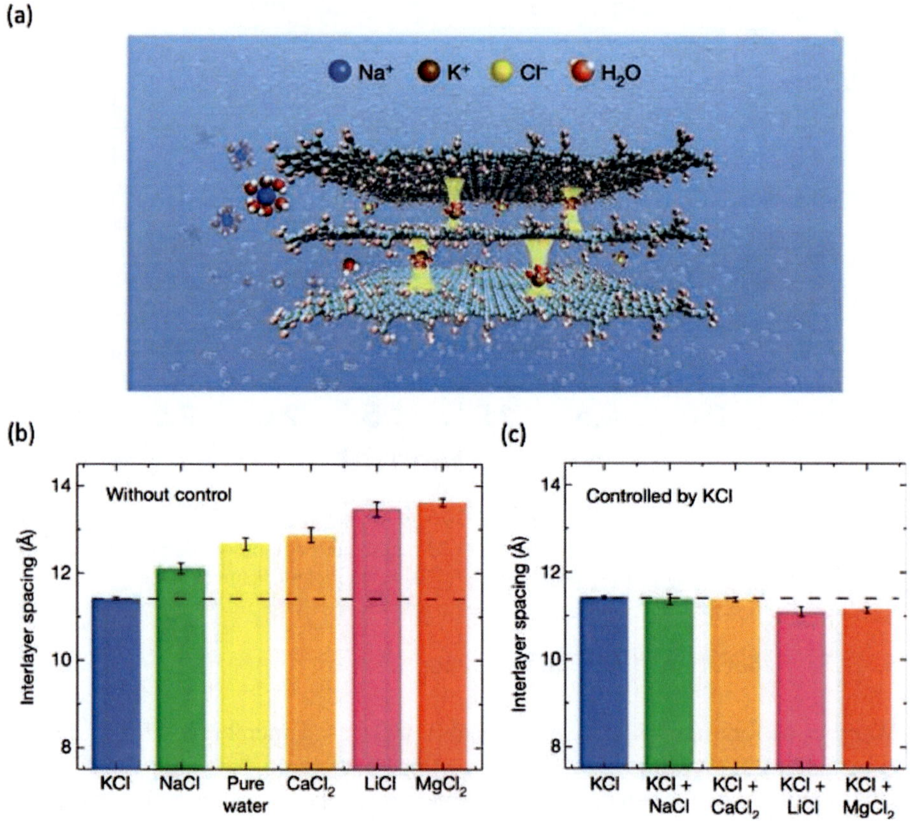

Fig. 5.35 **a** Schematic representation of how K^+ ions control the interlayer spacing of GO films. **b** Interlayer spacing for GOMs immersed in pure water or in various 0.25 mol l^{-1} (0.25 M) salt solutions. **c** Interlayer spacing of GOMs soaked in KCl solution, followed by immersion in various salt solutions. Error bars indicate the standard deviation from three different points of a sample. Reprinted by permission from Springer Nature [28], Copyright Springer Nature Singapore Pte Ltd. 2017

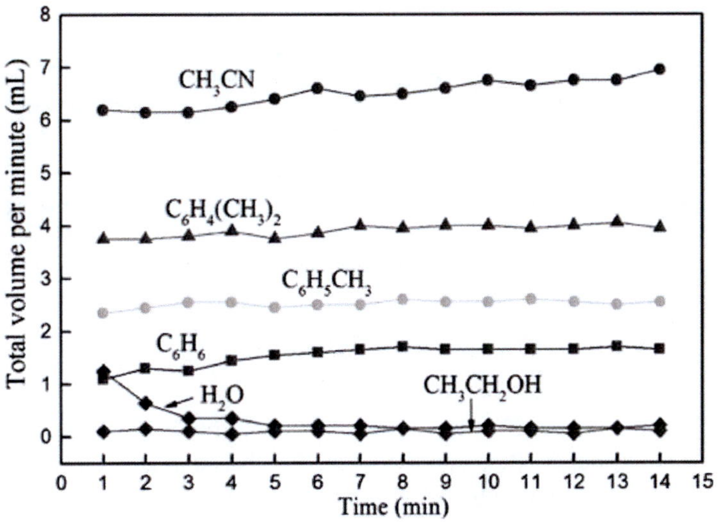

Fig. 5.36 Permeation behaviors of the M-GO membranes using NaCl. The *y*-axis represents the total volume of the solvent collected through the M-GO membrane each minute. The *x*-axis represents time (minute). From [29] Copyright © 2017 by John Wiley Sons, Inc. Reprinted by permission of John Wiley & Sons, Inc.

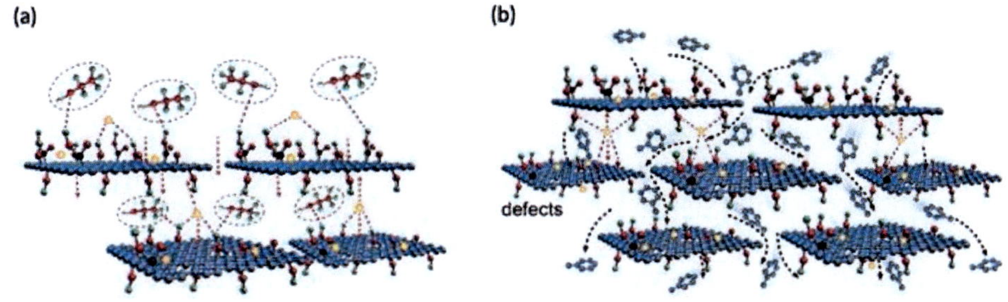

Fig. 5.37 **a** Interaction between ethanol molecules and the M-GO nanosheets. **b** The interaction between aromatic solvent molecules and the M-GO membrane. From [29] Copyright © 2017 by John Wiley Sons, Inc. Reprinted by permission of John Wiley & Sons, Inc.

Compared to these solvents, aromatic compounds have a slower rate of passage, which can be explained by cation–π interactions. Based on this selective permeability, C_2H_5OH/H_2O and C_2H_5OH/toluene mixtures can be effectively separated using cation-modified GO membranes.

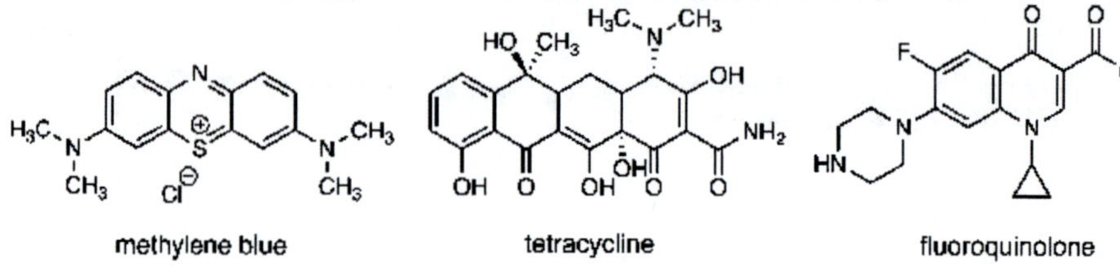

Fig. 5.38 Chemical structures of methylene blue, tetracycline, and fluoroquinolone

5.3.2 Adsorption of Cationic Molecules

GO is a promising and effective adsorbent for the removal of contaminants from aqueous environments. It is reported that GO-based materials are effective for the removal of aromatic contaminants such as dyes [30], tetracycline [31], fluoroquinolone antibiotics [32], cationic surfactants [33], and other organic pollutants [34]. It is suggested that the cation–π interactions between the positively charged functional groups and frameworks of these molecules (Fig. 5.38) with the π-electrons of GO contribute to the adsorption process as the major interaction.

5.3.3 Dispersion of Graphenes

Cation–π interactions are also involved in dispersion systems using graphene, which can be classified into the following three types: (1) dispersion of chemically modified graphene in aqueous solutions containing alkali metal ions, (2) dispersion of reduced graphene oxide (rGO) in ionic liquids and organic solvents, and (3) dispersion of GO in organic solvents and polymeric matrices by the surface modification of GO with ionic liquids.

5.3.3.1 Dispersion to Aqueous Solutions

High-quality rGOs with few defects or oxygen functional groups cannot be stably dispersed in a variety of solvents due to their hydrophobic nature. On the other hand, chemically modified graphenes (CMGs) can be stably dispersed in water by the interaction of graphene with monovalent cations without the use of dispersal agents. As shown in Fig. 5.39, stable rGO dispersions were obtained by the gentle reduction and aging of graphene oxide (GO) nanosheets dispersed in alkaline solvent [35].

Figure 5.39 shows that dispersion is accelerated with increases in the aging time to become constant at about 2 h. At this time, cation–π interactions between monovalent cations, such as Na^+ and K^+, and the aromatic ring on graphene are thought to have generated cation–π-interacting GO (CIGO) as shown in Fig. 5.39. The contribution

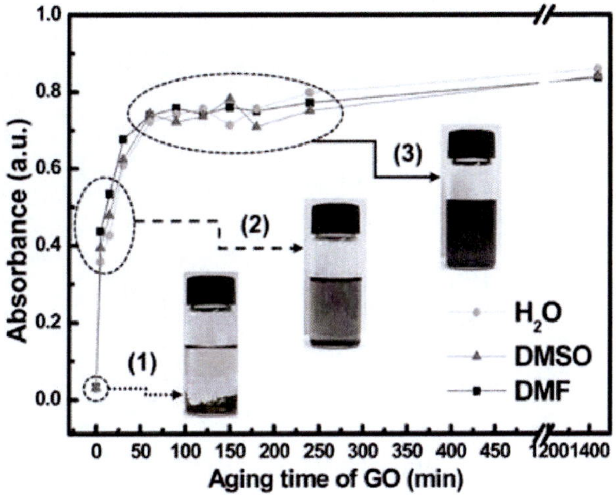

Fig. 5.39 Dispersion stability of CIRGO and non-interacting RGO. Absorbance as a function of aging for a GO solution in deionized water, DMSO or DMF. From [35] Copyright © 2012 by John Wiley Sons, Inc. Reprinted by permission of John Wiley & Sons, Inc.

of cation–π interactions is confirmed by the chemical shift of the sp^2 peak in the solid-state ^{13}C NMR spectra and Raman spectra.

Similar to graphene, the uniform dispersion of carbon nanotubes (CNTs) in aqueous solution or organic matrices remains a major challenge. Fukushima et al. found that when a suspension of single-walled carbon nanotubes (SWNTs) and an imidazolium ionic liquid, 1-butyl-3-methylimidazolium tetrafluoroborate $[bmim][BF_4]$ [36], was ground in an agate mortar and then centrifuged, a gel of pure single-walled carbon nanotubes was formed (Fig. 5.40). The heavily entangled bundles of nanotubes were found to unravel into thinner bundles in the gel. This is thought to be due to the cation–π interactions between the imidazolium ions and the π-electron network of the CNTs. It has been reported that multiwalled carbon nanotubes can also be stably dispersed in water by using small amounts of ionic liquids $[emim\text{-}NH_2][Br]$ and $[epy\text{-}NH_2][Br]$ (Fig. 5.40) [37].

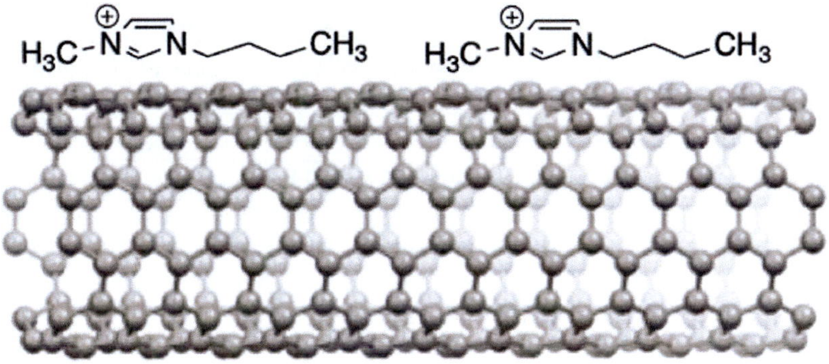

Fig. 5.40 Schematic representation of cation–π interactions between CNT and 1-butyl-3-methylimidazolium

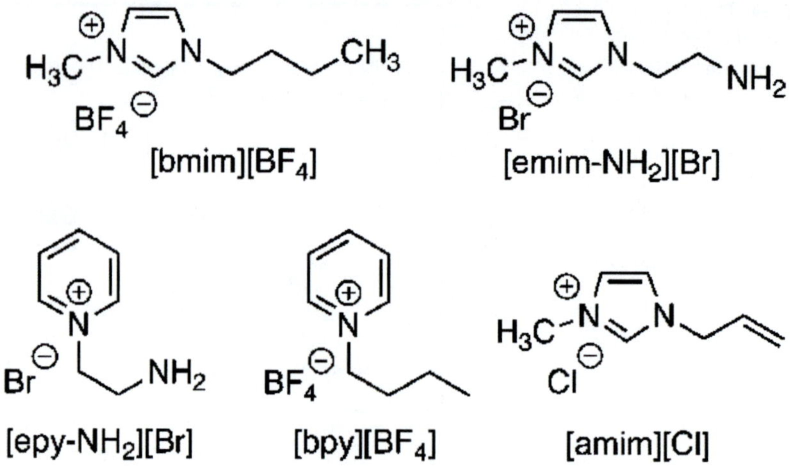

Fig. 5.41 Chemical structures of ionic liquids

5.3.3.2 Dispersion to Ionic Liquids

Ionic liquids are effective exfoliation media for the preparation of single layer graphenes due to their strong cation–π interactions with the π-electrons of graphite. Uniform and stable rGO dispersions have been obtained in ionic liquids at high concentrations without the help of dispersants [38]. Three hydrophilic ILs with 1-alkyl-3-methylimidazolium or N-alkylpyridinium cations, [bmim][BF_4], [amim][Cl] and [bpy][BF_4] shown in Fig. 5.41, were successfully used for stable dispersion at relatively high concentrations of rGO (up to 7.0 mg/ml).

5.3.3.3 Reversible Phase Transfer of Graphene Sheets

Kim et al. reported the reversible phase transfer of graphene between aqueous and organic solvents through the use of ionic liquid polymers (PILs) [39]. By changing the hydrophilic and hydrophobic properties of PILs, graphene can be dispersed in aqueous or organic layers, respectively.

Specifically, poly(1-vinyl-3-ethylimidazolium) salts were synthesized and exploited as functional materials that were capable of forming a complex with graphene sheets and switching their surface properties from a hydrophilic to a hydrophobic phase. The graphene sheets were reversibly transferred between aqueous and organic solutions through a simple PIL anion exchange treatment. When an equivalent of lithium bis(trifluoromethylsulfonyl) amide ($LiNTf_2$) was added to the aqueous suspension of PIL(Br)-G, black precipitates were formed in the water by the immediate substitution of the hydrophilic anions (Br) with hydrophobic anions (NTf_2) in the PIL. The resulting hydrophobic PIL(NTf_2)-modified graphene sheets,

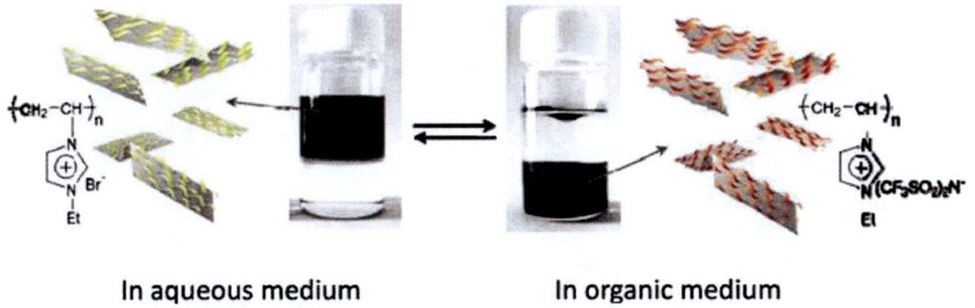

Fig. 5.42 Schematic illustration of the PIL-modified graphene sheets (PIL-G) showing the phase transfer of the PIL-modified graphene sheets between aqueous and organic solvent media by PIL anion exchange. Reprinted with permission from [39]. Copyright 2010 American Chemical Society

PIL(NTf_2)-G, were isolated and readily redispersed in a wide range of polar aprotic solvents (Fig. 5.42).

5.3.3.4 Dispersion to Polymers

Properties such as excellent mechanical strength and high electrical and thermal conductivity make graphene a promising building block for the preparation of high-performance polymer nanocomposites. The incorporation of graphene into polymers has been extensively studied in recent years. Ionic liquids are excellent dispersants of rGO and have also been reported as surface modifiers and dispersion enhancers for graphene in polymer matrices. As shown in Fig. 5.43, GO can be dispersed in polyurethane (PU) by its reduction with hydrazine and surface modification with ionic liquids [40]. The filler is uniformly dispersed in the polyurethane matrix due to a combination of cation–π and π–π interactions. This ionic liquid-modified graphene nanocomposite was used to improve the mechanical and thermal properties of PU.

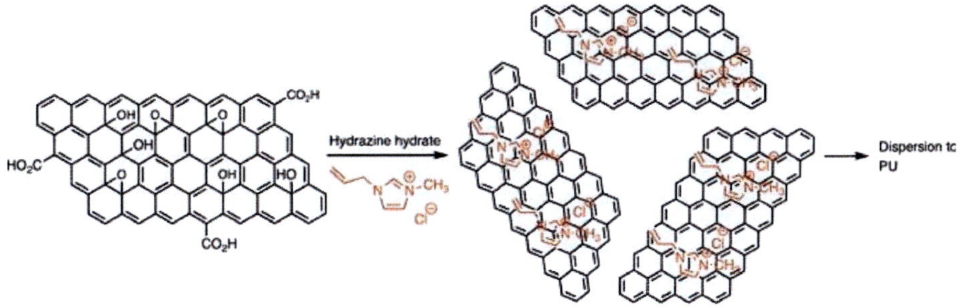

Fig. 5.43 Dispersion of rGO in IL by reducing GO with hydrazine, which can then be dispersed to polyurethane

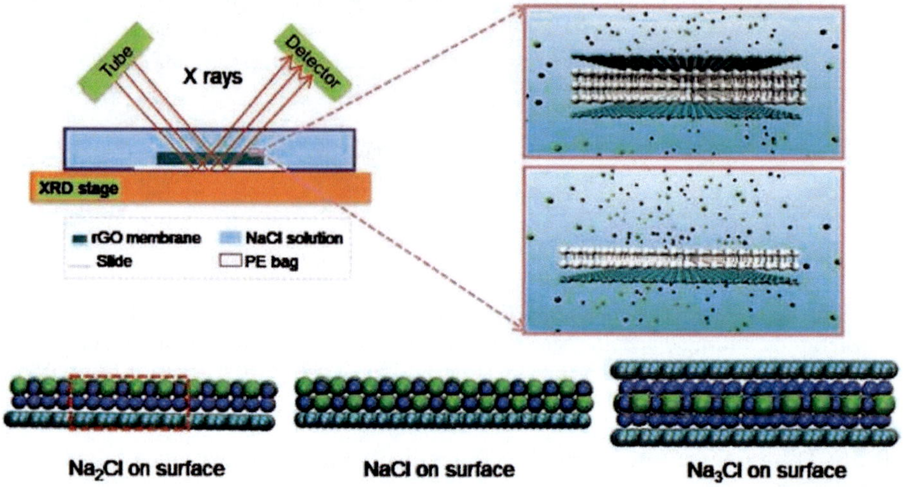

Fig. 5.44 Experimental setup for the preparation of 2D Na-Cl crystals on graphene surfaces. In the enlarged picture, the white lattices represent the 2D Na-Cl crystals. The cyan background represents the NaCl solution. Structure snapshots of Na_2Cl, NaCl, and Na_3Cl crystals on graphene sheets based on theoretical calculations. Cyan, blue, and green spheres represent C atoms, Na, and Cl, respectively. Reprinted by permission from Springer Nature [41], Copyright Springer Nature Singapore Pte Ltd. 2018

5.3.4 Formation of Non-Stoichiometric Na_xCl Crystals

When reduced graphene oxide (rGO) is immersed in a NaCl solution, the local concentration of sodium ions near the graphene surface increases due to the strong interaction between Na^+ and graphene. This not only affects the interfacial properties, but also the distribution of counter ions around the graphene in solution. As shown in Fig. 5.44, non-stoichiometric Na_2Cl and Na_3Cl crystals are precipitated on the surface of rGO film and natural graphite powder in NaCl solution [41]. These results strongly suggest that the unique distribution of ions on rGO through strong $Na^+\cdots\pi$ interactions leads to the creation of novel materials with unique stoichiometry, morphology, and other properties.

5.4 Energy-Related Materials

5.4.1 Energy Generation Systems

In 2014, Yin et al. found that moving a droplet of aqueous NaCl solution along a monolayer graphene strip generated a voltage [42]. As shown in Fig. 5.45, moving a 0.6 M NaCl solution at a constant velocity of 2.25 cm/s produced a voltage of about 0.15 mV. This is attributable to the accumulation of cations near the graphene

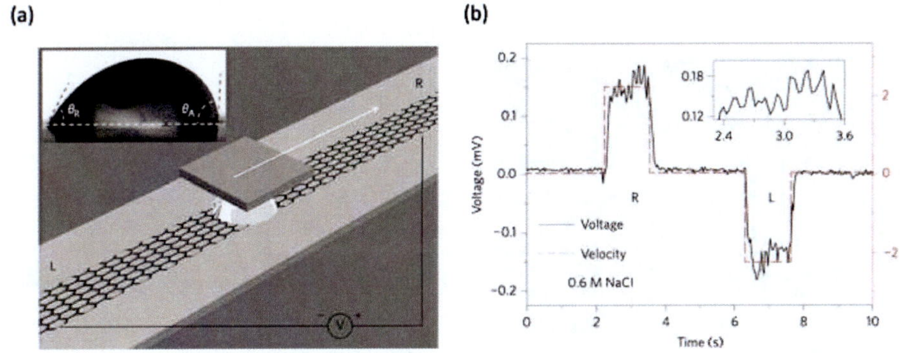

Fig. 5.45 **a** A liquid droplet is sandwiched between graphene and a SiO_2/Si wafer, and drawn by the wafer at specific velocities. Inset: a droplet of 0.6 M NaCl solution on a graphene surface with advancing and receding contact angles of ~91.98° and ~60.28°, respectively. **b** Typical voltage signal produced by drawing a droplet on a graphene strip from left (L) to right (R) and then back. The inset highlights the voltage signal during droplet movement. Reprinted by permission from Springer Nature [42], Copyright Springer Nature Singapore Pte Ltd. 2014

surface due to the strong interaction between cations and graphene, as described in Sect. 5.3.4.

Figure 5.46 shows the charge distribution and energy of hydrated sodium ions adsorbed on graphene as predicted by DFT calculations. It can be seen that the adsorption energy increases as the number of ions increases. Furthermore, when the ion droplet is pulled forward so that the droplet moves on the graphene, the ions are adsorbed at the front end, the pseudocapacitor moves forward, and electrons are pulled into the graphene. On the other hand, at the rear of the droplet, ions are desorbed and the pseudocapacitor discharges, releasing electrons into the graphene. This whole process results in an increase or decrease in electron density before and after the moving droplet compared to the static state, and the forward potential is higher than the backward potential. In this way, the experimental results can be explained by simulation (Fig. 5.46).

Such a new concept of energy generation using graphene has received considerable attention, and potential applications in various fields have been suggested. A highly efficient humidity-driven generator has also been developed based on cation–π interactions between ions and wrinkled graphene [43].

5.4.2 Energy Storage Systems

5.4.2.1 CNT-IL Systems

Carbon nanotubes (CNTs) have attracted much attention in many fields due to their excellent electrical, thermal, and mechanical properties. Nevertheless, they remain difficult to disperse, particularly in aqueous environments. Li et al. exploited

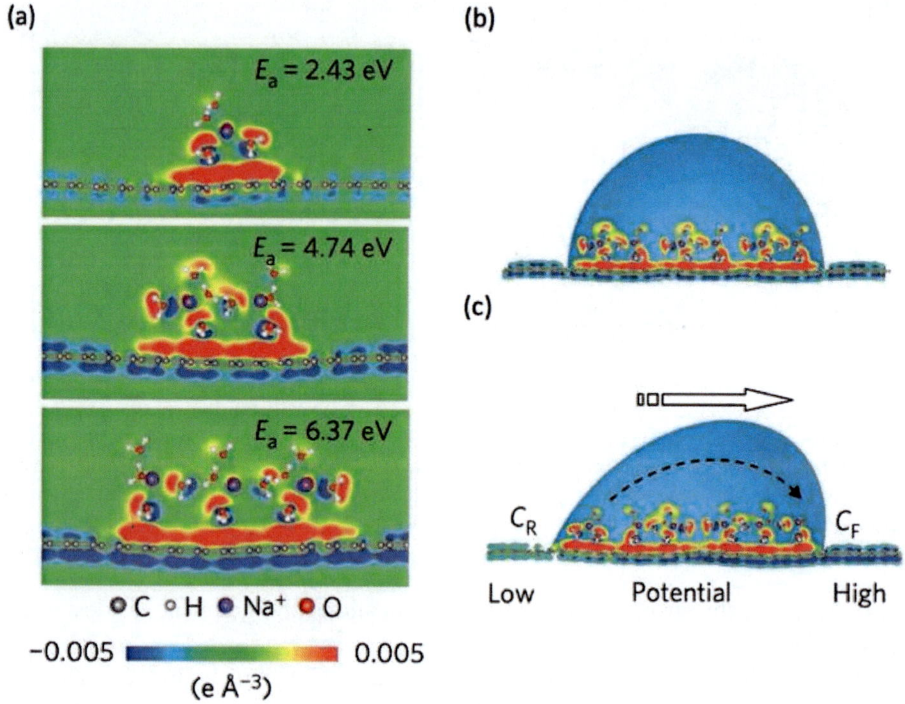

Fig. 5.46 **a** DFT results for the distribution of differential charge near a monolayer graphene strip due to the adsorption of one to three rows of hydrated sodium cations, and the corresponding adsorption energy (*E*a). **b** Schematic illustration of the pseudocapacitance formed by a static droplet on graphene. **c** Schematic illustration of the potential difference induced by a moving droplet. Reprinted by permission from Springer Nature [42], Copyright Springer Nature Singapore Pte Ltd. 2014

the cation–π interactions of imidazolium-based ionic liquid [Bmim][BF_4] to form conductive channels between single-walled carbon nanotubes (SWNTs) and iron fluoride nanoparticles [44]. Iron fluoride nanoparticles formed in situ (~10 nm) were obtained as micrometer-sized aggregates after wiring only 5 wt% carbon nanotubes. These composites acted as cathodes and showed a remarkable improvement in electrical capacity and rate performance (e.g., 220 mAh/g at 0.1 °C and 80 mAh/g at 10 °C).

The structure of the complex can be considered to resemble that shown in Fig. 5.47, where the SWNTs are uniformly modified by [Bmim][BF_4] molecules and the cations are oriented on the surface of the SWNTs. The ILs greatly weaken the cross-linking between the SWNTs, allowing the SWNTs to be well dispersed in the IL medium. On the other hand, the outwardly stretched BF_4 component reacts with the surrounding Fe(III) precursors to form iron-based fluoride nanoparticles.

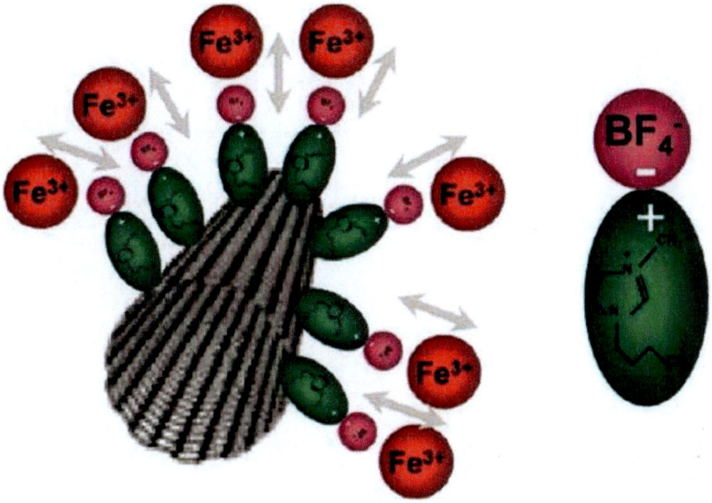

Fig. 5.47 Scheme of the respective interactions of [Bmim][BF_4] with SWT and the $Fe(NO_3)_39H_2O$ precursor. Reprinted with permission from [44]. Copyright 2011 American Chemical Society

5.4.2.2 Organic Anode Materials

Rechargeable lithium-ion batteries (LIBs) are widely used in electric vehicles, wearable electronic devices, and portable consumer electronics and have attracted much interest as energy storage systems. Currently, inorganic active materials are mainly employed as anode active materials in LIBs. However, these inorganic anode materials have drawbacks such as toxicity, non-renewability, and contaminant release. On the other hand, organic electrode materials are promising alternatives that can overcome the disadvantages of metal-based inorganic electrodes. However, due to the low solubility, low capacity, and poor conductivity of these organic electrode materials, they have not yet been used as anode materials for rechargeable batteries.

Wei et al. prepared an indole-based conjugated microporous polymer, poly(bisindolylmaleimide) (PBIM), by oxidative coupling polymerization and showed that it can be used as an anode material for lithium-ion batteries with excellent electrochemical performance (Fig. 5.48a and b) [45]. Due to the unique aromatic structure containing heteroatoms and the reversible cation–π interactions involved, the batteries with PBIM anodes exhibit excellent cycling stability (1000 cycles) and excellent rate performance at high capacity (1172 mA h/g, current density 50 mA/g). The high specific capacity and outstanding cycle stability of PBIMs are thought to result from the cation–π interactions between Li^+ and the π-face of the indole ring, which shortens the Li^+ diffusion path and improves the reaction rate and specific capacity of the battery. Molecular dynamics (MD) and density functional theory (DFT) calculations reveal a cation–π interaction (~3.11 Å) between Li^+ and indole π-plane (Fig. 5.48c).

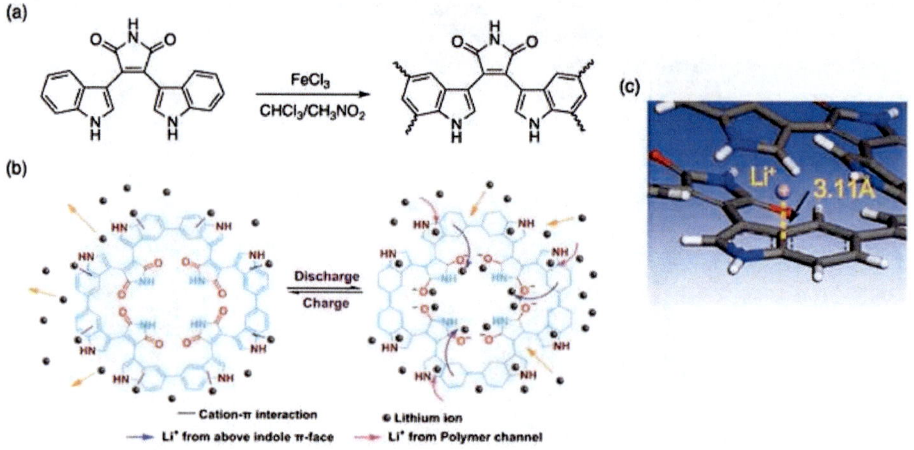

Fig. 5.48 **a** Synthetic route to PBIM and **b** reaction mechanisms. **c** The polymer and Li^+ in a simulation box under an equilibrium state. Reproduced from Ref. [45] with permission from the Royal Society of Chemistry

Yang et al. prepared a different type of π-conjugated polymer, poly(triazatruxene) (PTAT), using an indole derivative as shown in Fig. 5.49 [46]. The PTAT was pretreated by immersion in lithium chloride solution to yield PTAT@LiCl (Fig. 5.50). The cation–π (Li^+-π) interaction provides space for a large number of lithium ions to be accommodated, which then undergoes a lithiation reaction with the active groups on the electrode surface. This study demonstrates the great potential of lithium-doped π-conjugated polymers as high-performance anodes.

Raman measurements were performed on both PTAT@LiCl and the initial PTAT, and the spectra were compared (Fig. 5.51). There was a clear change in Raman intensity and displacement at 897 cm^{-1}, 1220 cm^{-1}, and 1414 cm^{-1}, reflecting the cation–π interactions between the indole group and Li ions.

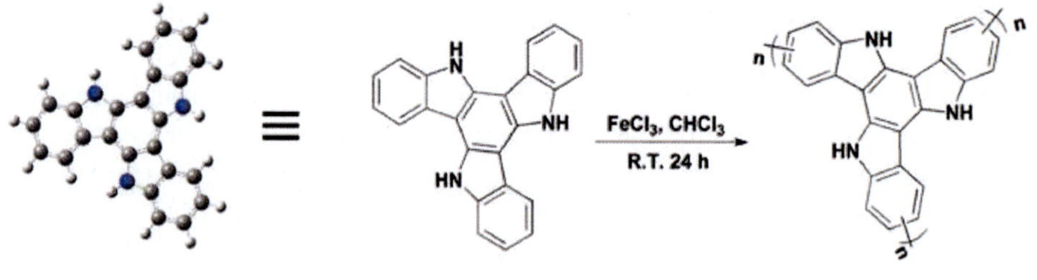

Fig. 5.49 Synthetic routes to PTAT. Reprinted from [46], Copyright 2020, with permission from Elsevier

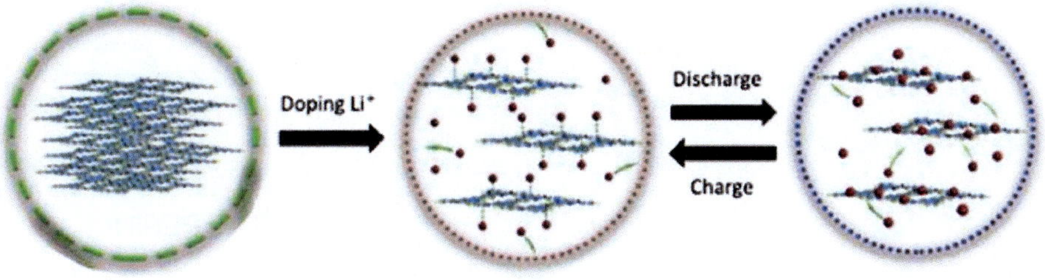

Fig. 5.50 Lithium storage mechanisms. Reprinted from [46], Copyright 2020, with permission from Elsevier

Fig. 5.51 Raman spectra of PTAT@Li and PTAT. Reprinted from [46], Copyright 2020, with permission from Elsevier

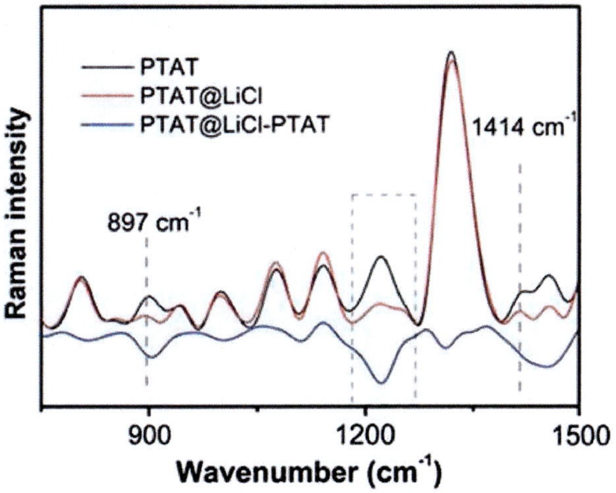

5.5 Fluorophores

5.5.1 Fluorescent-Switch Polymers

The use of polymers that allow switchable fluorescence has recently shown promise as an attractive approach in environmental or biological fields. Chang et al. found that the on/off switching of fluorescence emissions of a polymer (N-arylene indole ether sulfone) (PESIN) (Fig. 5.52) with an indole derivative can be controlled by cation–π interactions [47]. Adjustment of the protonation/deprotonation state of the pyridines affects the formation of the cation–π interactions, leading to the on/off switching of fluorescence emissions.

Figure 5.53a, b shows UV and fluorescence spectra of PESIN and fluorescence images of PESIN in DMF, respectively. Addition of Py·HCl to a PESIN solution caused fluorescence quenching at 458 nm, with the fluorescence intensity of PESIN decreasing with increases in the pyridine hydrochloride concentration (Fig. 5.53c).

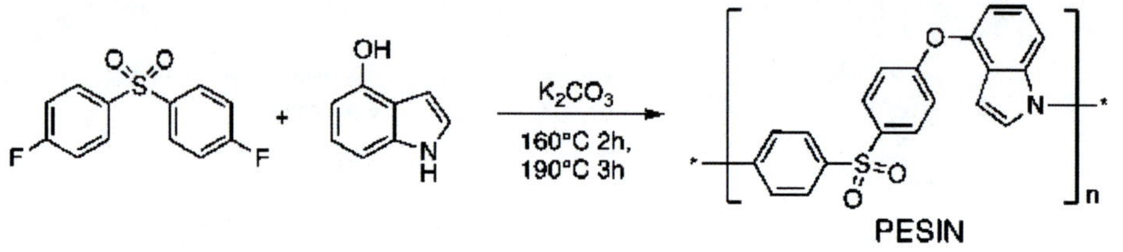

Fig. 5.52 Synthesis of the polymer sensor PESIN

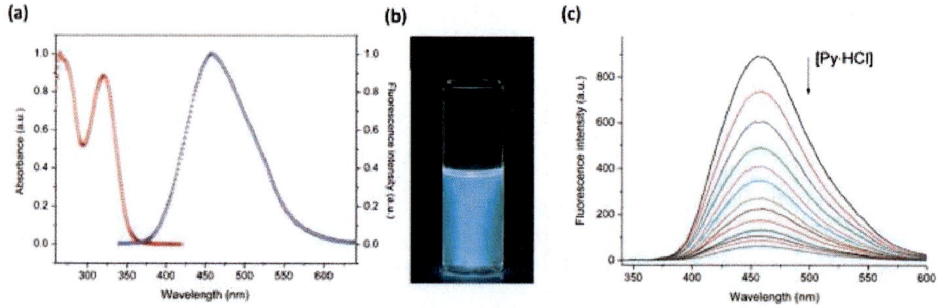

Fig. 5.53 **a** UV and fluorescence spectra of PESIN in DMF solution (10 μM, $\lambda_{ex} = 320$ nm). **b** Fluorescence images of PESIN in DMF ($\lambda_{ex} = 365$ nm). **c** Fluorescence spectra ($\lambda_{ex} = 320$ nm) of the DMF–H_2O (99:1, v/v) solution of PESIN (10 μM) in the presence of Py·HCl (From top to bottom: 0, 25, 50, 75, 100, 125, 150, 175, 200, 225, 250, 275, 300 equiv.) at 25 °C. Reproduced from Ref. [47] with permission from the Royal Society of Chemistry

This suggests the interruption of intramolecular charge transfer by a chelation-enhanced quenching (CHEQ) effect with Py·HCl through cation–π interactions. Molecular dynamics simulation calculations show that the distance between the -NH^+···indole ring is about 3.21 Å, supporting the existence of a strong cation–π interaction between these groups.

Based on the above results, a mechanistic pathway for the formation of the complex between PESIN and pyridine and HCl is proposed in Fig. 5.54. PESIN emits a cyan light in solution (Fig. 5.54a), and the addition of HCl or pyridine has little effect on the fluorescence intensity (Fig. 5.54b). In contrast, the addition of pyridine or HCl to the solution leads to the formation of Py·HCl in the system, which causes the fluorescence to disappear on interaction with PESIN (Fig. 5.54d). Furthermore, as shown in Fig. 5.54e, the introduction of basic OH-anions into the quenched solution causes the deprotonation of pyridine hydrochloride, and the fluorescence reappears. It is suggested that the ON—OFF response is mainly governed by the generation and disappearance of the cation–π interaction.

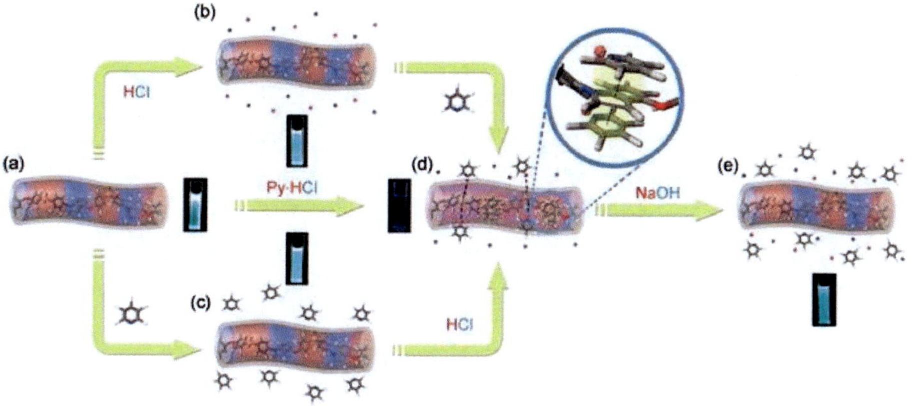

Fig. 5.54 Schematic representation of the mechanism by which the polymer sensor PESIN can recognize Py·HCl in an aqueous solution. Reproduced from Ref. [47] with permission from the Royal Society of Chemistry

5.5.2 *HaloTag Fluorogenic Probes*

Liu et al. developed a new type of fluorescence-generating probe **8** for HaloTag (Fig. 5.55) with a benzothiadiazole structure and which exhibits 1000-fold fluorescence enhancement upon reaction with HaloTag [48]. To understand the mechanism of fluorescence enhancement, the cocrystal structure of the complex of HaloTag substrate (**8**) and the protein was determined (PDB: 5UXZ). The benzothiadiazole moiety of **8** is located at the binding site of the Halo ligand-binding tunnel and interacts directly with tryptophan 141 (Fig. 5.56a). The N atom of the dimethylamino group is located near the geometrical center of the benzene ring.

When the W141A mutation is used, the fluorescence of the **8**-Halo complex is completely eliminated, clearly showing the importance of Trp to the luminescence. Furthermore, mutation of the Trp residue to 5-fluoro-L-Trp (5FW) greatly reduces the

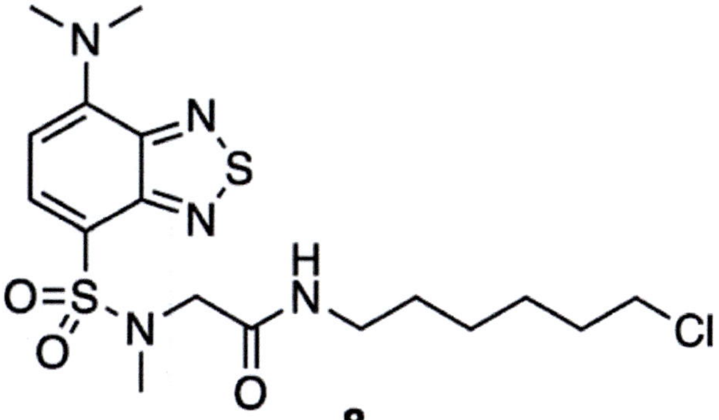

8

Fig. 5.55 Chemical structure of the HaloTag substrate **8**

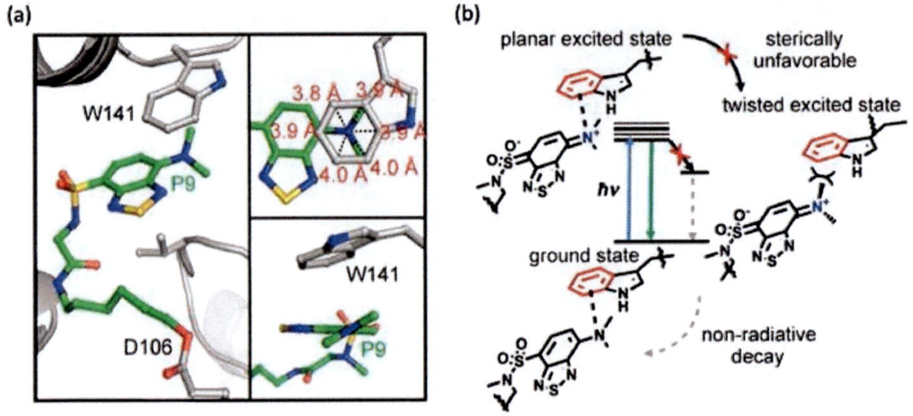

Fig. 5.56 **a** Structure of the cocrystal of the HaloTag substrate **8** and protein, showing close contact of the Me_2N group and the benzene ring of Trp141. **b** Schematic representation of the ground state and planar and twisted excited states of an **8**-Trp conjugate in the protein. The Me_2N group of **8** is thought to take on a positive charge in the excited state and interact with the aromatic group in Trp141 via a cation–π interaction. Reprinted with permission from [48]. Copyright 2017 American Chemical Society

fluorescence intensity of the HaloTag-**8** complex. This is attributed to the weakened cation–π interaction in 5FW due to its lower electron density than that of Trp.

Theoretical calculations show that the *N,N*-dimethylamino group of **8** is almost neutral in the ground state, but has a positive charge in the excited state. These suggest that the photoexcited state of benzothiadiazole can be stabilized by Trp through a cation–π interaction. This stabilization allows the excited chromophore to be better retained and minimizes the radiation-free decay via the twisted intramolecular charge-transfer mechanism (Fig. 5.56b). As a result, substrate **8** bound to HaloTags exhibits high fluorescence intensity and a long fluorescence lifetime. In fact, time-resolved fluorescence measurements revealed that the fluorescence lifetime of the HaloTag-**8** complex is 100 times longer than that of **8** alone. This study suggests that the cation–π interaction in the excited state is an effective tool with which to enhance the optical performance of fluorescent sensors.

5.5.3 Sensors for Alkyl Halides

Several pyridinium compounds have been used as emission luminogens in both solution and solid states. Chen et al. reported that simple *N*-methylpyridinium derivatives **9a–9c** function as luminogens [49]. The luminescence intensity depends on the chain length of the compounds, with **9b** showing the strongest luminescence (Fig. 5.57a). As the closed conformer of **9b** is more stabilized than those of **9a** and **9c** due to the intramolecular cation–π interaction between the pyridinium and benzene rings, the interaction is thought to be responsible for the luminescent intensity.

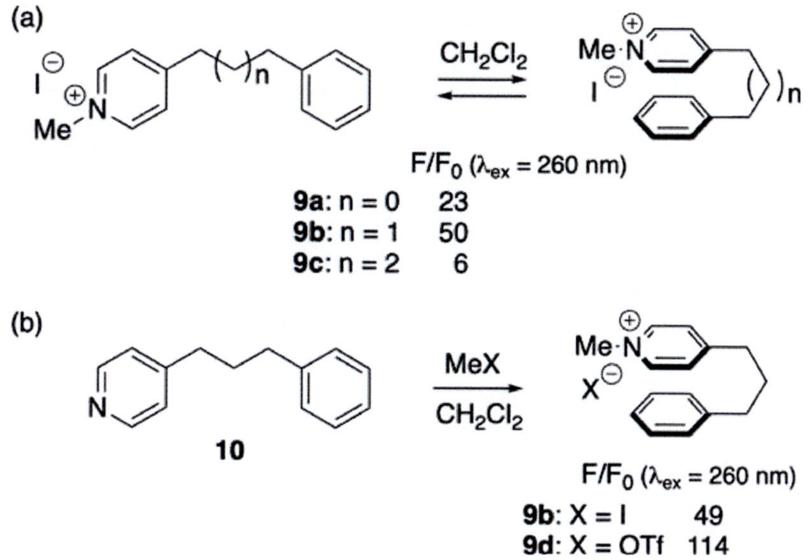

Fig. 5.57 **a** Screening of the fluorescent responses of N-alkylpyridinium salts **9a–9c**. **b** Fluorescence response by methylation of 10 as a sensor

This phenomenon has been used for sensing the presence of alkylating agents in solution by 1-(4-pyridyl)-3-phenylpropane (**10**). The reactions of the pyridine ring of **10** with iodomethane and methyl triflate produce *N*-methylpyridinium salts **9b** and **9d**, respectively, with the luminescence emitted from which signaling the presence of the alkylating agents (Fig. 5.57b).

5.5.4 Aggregation-Induced Emission Luminogens

Leduskrasts et al. developed a solid-state fluorophore **11b** that contains pyridinium and carbazole moieties (Fig. 5.58a) [50]. Figure 5.58b shows two neighboring molecules extracted from the crystal packing diagrams of **11a** and **11b**. The centroid distances between the pyridinium rings and the benzene rings of carbazole of **11a** and **11b** are 3.500 Å, 3.652 Å, and 3.525 Å, respectively, indicating that cation–π interactions contribute to the formation of dimeric structures. Although the crystal structures of **11a** and **11b** are very similar to each other, as described above, their luminescence properties are quite different: The photoluminescence quantum yield (ϕ) of **11b** in the solid state is very high (ϕ= 79.8% at λ = 489 nm), while a much lower intensity is observed for **11a**. This may be due to the difference in the counter anions, where the iodide ion quenches the emission more effectively than does the perchlorate ion.

In addition, these compounds do not emit light in MeOH solution, indicating that fluorophore **11b** has an aggregation-induced emission property. The TD-DFT

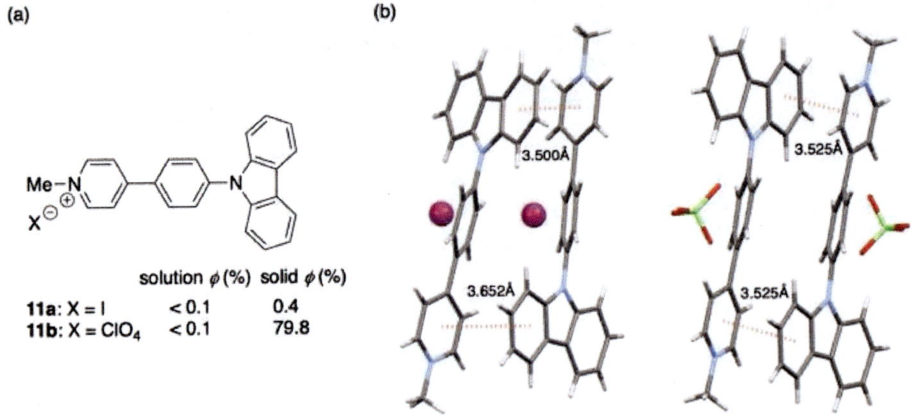

Fig. 5.58 **a** Chemical structures and **b** crystal structures of pyridinium salts **11a** and **11b**

calculations for cation–π-controlled crystals of **11b** suggest that the observed solid-state luminescence properties arise from intermolecular charge transfer between the pyridinium and benzene rings.

5.5.5 Intercalation-Induced Emission Systems

Hiratani and Konishi found that nanoparticles of CdS, a group 12 metal chalcogenide compound, have characteristic size-dependent optical properties [51]. $Cd_{10}S_4(SPh)_{12}$ (**12**) dissolves in chloroform in the presence of tetraoctylammonium bromide (TOAB), the emission spectrum of which shows a maximum wavelength at 605 nm. When cetyltrimethylammonium bromide (CTAB) is added, the maximum wavelength shifts to a longer wavelength at 617 nm, and the emission intensity increases 8.5-fold, while there is little change in the absorption spectrum (Fig. 5.59).

This is thought to be due to the intercalation of CTA^+ between the phenyl groups of **12** by cation–π interactions. Job's plots of CTAB and cetylpyridinium bromide (CPB) on the **12**/TOAB complex showed that 12 cations bind to the 12 Ph groups of **12** to form corresponding complexes. The association constants were 94 M^{-1} for CPB and 70 M^{-1} for CTAB.

Zhong et al. found that when a cationic surfactant, cetyltrimethylammonium bromide (CTAB), and a fluorescent polycyclic aromatic hydrocarbon are cointercalated into the clay, strong luminescence occurs within the laminated clay due to the cation–π interaction [52]. By using this phenomenon, they succeeded in the 3D imaging of the macrodispersion of clay filler in the silicone rubber matrix using confocal laser scanning microscopy (CLSM).

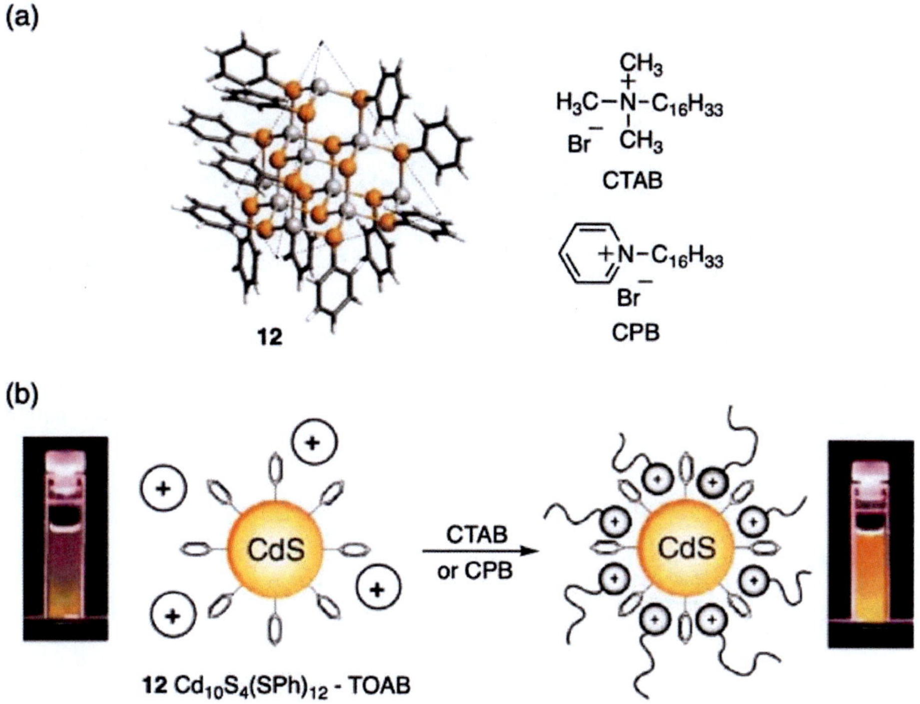

Fig. 5.59 **a** Structure of **12** and **b** change in the fluorescence color by intercalation of CTA^+ between the phenyl groups of **12** by cation–π interactions. From [51] Copyright © 2004 by John Wiley Sons, Inc. Reprinted by permission of John Wiley & Sons, Inc.

5.6 Recognition-Based Materials

5.6.1 Cavitand Receptors for AFM Prove Recognition

The recognition of ammonium by cavitand receptors through cation–π interactions is well known, and Eckel et al. have used this interaction for atomic force microscopy [53]. The formation of complexes of 2,8,14,20-tetra-(10-(decylthio)decyl)cavitand (**13**) with ethylammonium and trimethylethylammonium has been predicted by calculations (Fig. 5.60). Based on this prediction, they used embedded cavitand **13** as a receptor molecule on a gold surface modified with didecylsulfide, and unimolecular ammonium or trimethylammonium was used as a probe for atomic force microscopy (AFM) linking to the end of a PEG linker. As expected, it was found that the probe approaches the receptor due to the cation–π interaction between the single molecule of the cavitand and the cation (Fig. 5.61). The interaction force was reported to be −28 kJ/mol for ammonium and −38 kJ/mol for trimethylammonium.

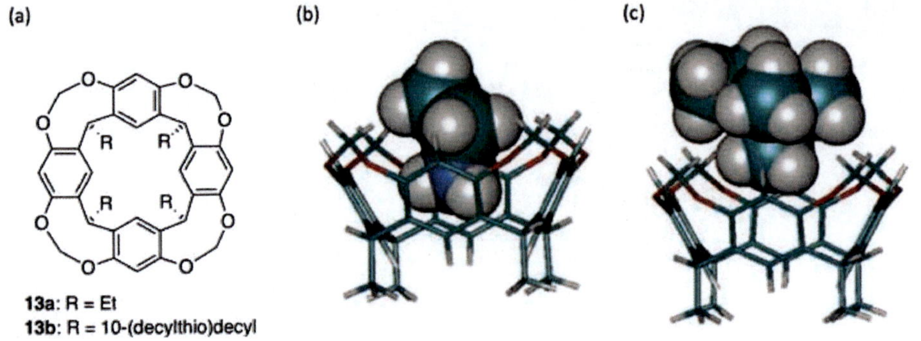

Fig. 5.60 **a** Chemical structures of **13a** and **13b**, and gas-phase structures for the complexes between **13a** and **b** ethylammonium ion and **c** ethyltrimethylammonium ions. Structure optimized at the B3LPY/3-21G* level. From [53] Copyright © 2005 by John Wiley Sons, Inc. Reprinted by permission of John Wiley & Sons, Inc.

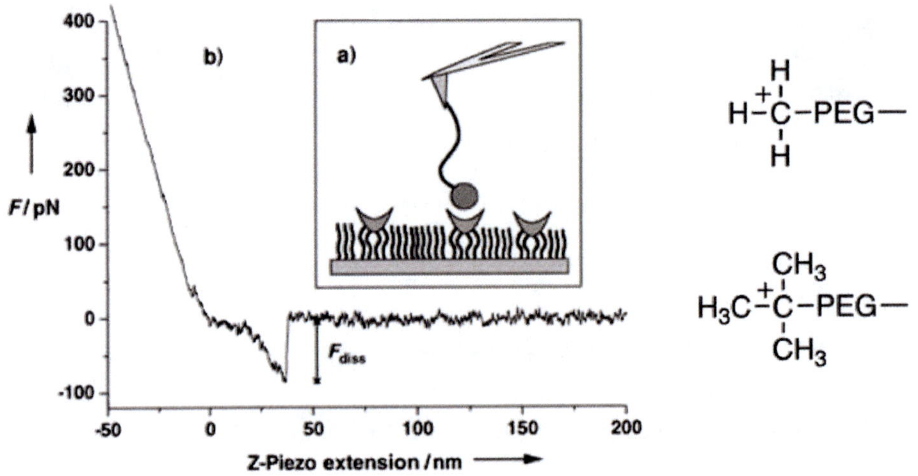

Fig. 5.61 Force spectroscopy: **a** Schematic setup in which the cavitand is immobilized together with didecylsulfide in a 1:40 mixture on a gold substrate. The (tetraorganyl) ammonium residue (shaded circle) is attached to a Si_3N_4 AFM tip by a flexible polymer linker. **b** Typical force–distance curve. The stretching of the PEG linker over a certain distance prior to bond rupture (tip detachment and relaxation of the cantilever) indicates an unbinding event. From [53] Copyright © 2005 by John Wiley Sons, Inc. Reprinted by permission of John Wiley & Sons, Inc.

5.6.2 Sumanene–Ferrocene Conjugates for Cs^+ Recognition

Kasprzak and Sakurai have developed a sumanene–ferrocene probe that efficiently and selectively recognizes a cesium ion [54]. Inspired by novel organometallic sumanene derivatives for Cs^+ recognition synthesized by Rogachev's group [55],

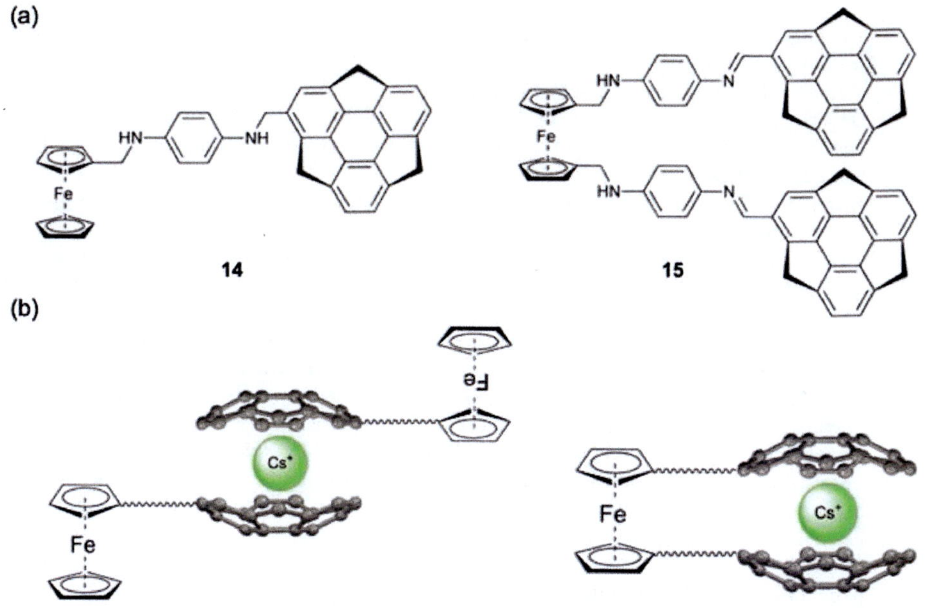

Fig. 5.62 **a** Structures of ferrocenyl derivatives of sumanene **14** and **15**. **b** Hypothetical structures for the complexes of Cs^+ with sumanene **14** and sumanene **15**. Reproduced from Ref. [54] with permission from the Royal Society of Chemistry

they hypothesized that ferrocenyl derivatives of sumanene could function as effective Cs^+ recognition materials and synthesized new conjugates of ferrocene and sumanene **14** and **15** (Fig. 5.62a).

The properties of the complexes were investigated by fluorescence spectroscopy; the titration experiments for Cs^+ and compounds **14** and **15** showed a decrease in the emission intensity. Furthermore, Job's plots revealed that the di-sumanenyl ferrocene derivative **15** forms a 1:1 complex with Cs^+, while the monosumanenyl compound **14** forms a 2:1 complex (Fig. 5.62b). These results indicate that the site-selective cation–π interaction between Cs^+ and the concave surface of the sumanene was the driving force behind the recognition of Cs^+.

References

1. Freire F, Seco JM, Quinoa E, Riguera R (2011) Chiral amplification and helical-sense tuning by mono- and divalent metals on dynamic helical polymers. Angew Chem Int Ed 50:11692–11696
2. Arias S, Freire F, Quiñoá E, Riguera R (2015) The leading role of cation–π interactions in polymer chemistry: the control of the helical sense in solution. Polymer Chem 6:4725–4733
3. Abe H, Masuda N, Waki M, Inouye M (2005) Regulation of saccharide binding with basic poly(ethynylpyridine)s by H^+-induced helix formation. J Am Chem Soc 127:16189–16196
4. Heemstra JM, Moore JS (2004) Folding-promoted methylation of a helical DMAP analogue. J Am Chem Soc 126:1648–1649

5. Smaldone RA, Moore JS (2007) Foldamers as reactive sieves reactivity as a probe of conformational flexibility. J Am Chem Soc 129:5444–5450
6. Yamada S (2018) Cation–π interactions in organic synthesis. Chem Rev 118:11353–11432
7. Dolain C, Zhan C, Léger J-M, Daniels L, Huc I (2005) Folding directed N-oxidation of oligopyridine-dicarboxamide strands and hybridization of oxidized oligomers. J Am Chem Soc 127:2400–2401
8. Aydogan A, Sessler JL (2014) An imidazolium-functionalized self-assembling calix[4]pyrrole. Chem Commun 50:13600–13603
9. Yuvayapan S, Aydogan A (2019) Counter cation dependent and stimuli responsive supramolecular polymers constructed by calix[4]pyrrole based host-guest interactions. Eur J Org Chem 633–639.
10. Chang G, Yang L, Yang J, Stoykovich MP, Deng X, Cui J, Wang D (2018) High-performance pH-switchable supramolecular thermosets via cation–π interactions. Adv Mater 30:1704234
11. Yang L, Du M, Yang S, Wei W, Chang G (2018) Recyclable crosslinked high-performance polymers via adjusting intermolecular cation–π interactions and the visual detection of tensile strength and glass transition temperature. Macromol Rapid Commun 39:1800031
12. Shi L, Zhang W, Yang K, Shi H, Li D, Liu J et al (2015) Antibacterial and osteoinductive capability of orthopedic materials via cation–π interaction mediated positive charge. J Mater Chem B 3:733–737
13. Wang R, Xie T (2010) Macroscopic evidence of strong cation-π interactions in a synthetic polymer system. Chem Commun 46:1341–1343
14. Fan H, Wang J, Tao Z, Huang J, Rao P, Kurokawa T et al (2019) Adjacent cationic-aromatic sequences yield strong electrostatic adhesion of hydrogels in seawater. Nat Commun 10:5127
15. Chang L, Liu H, Ding Y, Zhang J, Li L, Zhang X, Liu M, Jiang L (2017) A smart surface with switchable wettability by an ionic liquid. Nanoscale 9:5822–5827
16. Chang G, Wang Y, Wang C, Li Y, Xu Y, Yang L (2018) A recyclable hydroxyl functionalized polyindole hydrogel for sodium hydroxide extraction via the synergistic effect of cation-π interactions and hydrogen bonding. Chem Commun 54:9785–9788
17. Park S, Kim S, Jho Y, Hwang DS (2019) Cation-π interactions and their contribution to mussel underwater adhesion studied using a surface forces apparatus: a mini-review. Langmuir 35:16002–16012
18. Gebbie MA, Wei W, Schrader AM, Cristiani TR, Dobbs HA, Idso M, Chmelka BF, Waite JH, Israelachvili JN (2017) Tuning underwater adhesion with cation-π interactions. Nature Chem 9:473–479
19. Lu Q, Oh DX, Lee Y, Jho Y, Hwang DS, Zeng H (2013) Nanomechanics of cation-π interactions in aqueous solution. Angew Chem Int Ed 52:3944–3948
20. Kim S, Yoo HY, Huang J, Lee Y, Park S, Park Y et al (2017) Salt triggers the simple coacervation of an underwater adhesive when cations meet aromatic π electrons in seawater. ACS Nano 11:6764–6772
21. Lim C, Huang J, Kim S, Lee H, Zeng H, Hwang DS (2016) Nanomechanics of poly(catecholamine) coatings in aqueous solutions. Angew Chem Int Ed 55:3342–3346
22. Xie L, Gong L, Zhang J, Han L, Xiang L, Chen J et al (2019) A wet adhesion strategy via synergistic cation–π and hydrogen bonding interactions of antifouling zwitterions and mussel-inspired binding moieties. J Mater Chem A 7(38):21944–21952
23. Pillai KV, Renneckar S (2009) (2009) Cation-π interactions as a mechanism in technical lignin adsorption to cationic surfaces. Biomacromol 10:798–804
24. Ouyang X, Deng Y, Qian Y, Zhang P, Qiu X (2011) Adsorption characteristics of lignosulfonates in salt-free and salt-added aqueous solutions. Biomacromol 12:3313–3320
25. Zhao G, Zhu H (2020) Cation-π interactions in graphene-containing systems for water treatment and beyond. Adv Mater 32:1905756
26. Sun P, Zheng F, Zhu M, Song Z, Wang K, Zhong M, Wu D et al (2014) Selective trans-membrane transport of alkali and alkaline earth cations through graphene oxide membranes based on cation-π interactions. ACS Nano 8:850–859

27. Sun P, Liu H, Wang K, Zhong M, Wu D, Zhu H (2014) Selective ion transport through functionalized graphene membranes based on delicate ion-graphene interactions. J Phys Chem C 118:19396–19401
28. Chen L, Shi G, Shen J, Peng B, Zhang B, Wang Y et al (2017) Ion sieving in graphene oxide membranes via cationic control of interlayer spacing. Nature 550:380–383
29. Long Y, Wang K, Xiang G, Song K, Zhou G, Wang X (2017) Molecule channels directed by cation-decorated graphene oxide nanosheets and their application as membrane reactors. Adv Mater 29:1606093
30. Yan H, Tao X, Yang Z, Li K, Yang H, Li A et al (2014) Effects of the oxidation degree of graphene oxide on the adsorption of methylene blue. J Hazard Mater 268:191–198
31. Gao Y, Li Y, Zhang L, Huang H, Hu J, Shah SM et al (2012) Adsorption and removal of tetracycline antibiotics from aqueous solution by graphene oxide. J Colloid Interface Sci 368(1):540–546
32. Zhao Q, Zhang S, Zhang X, Lei L, Ma W, Ma C et al (2017) Cation-π interaction: a key force for sorption of fluoroquinolone antibiotics on pyrogenic carbonaceous materials. Environ Sci Technol 51:13659–13667
33. de Figueiredo NT, Kushima AP, Rodrigues SL, Bardelin NW, Prediger P (2020) Influence of adsorbent/adsorbate interactions on the removal of cationic surfactants from water by graphene oxide. Water Air Soil Pollut 231:304
34. Wang Y, Lyu L, Wang D, Yu H-Q, Li T, Gao Y, et al. (2021) Cation-π induced surface cleavage of organic pollutants with OH formation from H_2O for water treatment. iScience 24:102874.
35. Jeong SY, Kim SH, Han JT, Jeong HJ, Jeong SY, Lee G-W (2012) Highly concentrated and conductive reduced graphene oxide nanosheets by monovalent cation-π interaction: toward printed electronics. Adv Funct Mater 22:3307–1334
36. Fukushima T, Kosaka A, Ishimura Y, Yamamoto T, Takigawa T, Ishii N, Aida T (2003) Molecular ordering of organic molten salts triggered by single-walled carbon nanotubes. Science 300:2071–2074
37. Zhou X, Wu T, Ding K, Hu B, Hou M, Han B (2009) The dispersion of carbon nanotubes in water with the aid of very small amounts of ionic liquid. Chem Commun 1897–1899.
38. Zhang B, Ning W, Zhang J, Qiao X, Zhang J, He J et al (2010) Stable dispersions of reduced graphene oxide in ionic liquids. J Mater Chem 20:5401–5403
39. Kim T, Lee H, Kim J, Suh KS (2010) Synthesis of phase transferable graphene sheets using ionic liquid polymers. ACS Nano 4:1612–1618
40. Ma W-S, Wu L, Yang F, Wang S-F (2014) Non-covalently modified reduced graphene oxide/polyurethane nanocomposites with good mechanical and thermal properties. J Mater Sci 49:562–571
41. Shi G, Chen L, Yang Y, Li D, Qian Z, Liang S et al (2018) Two-dimensional Na-Cl crystals of unconventional stoichiometries on graphene surface from dilute solution at ambient conditions. Nat Chem 10:776–779
42. Yin J, Li X, Yu J, Zhang Z, Zhou J, Guo W (2014) Generating electricity by moving a droplet of ionic liquid along graphene. Nat Nanotechnol 9:378–383
43. Zhen Z, Li Z, Zhao X, Zhong Y, Huang M, Zhu H (2019) A non-covalent cation-π interaction-based humidity-driven electric nanogenerator prepared with salt decorated wrinkled graphene. Nano Energy 62:189–196
44. Li C, Gu L, Tong J, Maier J (2011) Carbon nanotube wiring of electrodes for high-rate lithium batteries using an imidazolium-based ionic liquid precursor as dispersant and binder: a case study on iron fluoride nanoparticles. ACS Nano 5:2930
45. Wei W, Chang G, Xu Y, Yang L (2018) An indole-based conjugated microporous polymer: a new and stable lithium storage anode with high capacity and long life induced by cation–π interactions and a N-rich aromatic structure. J Mater Chem A 6:18794–18798
46. Yang L, Wei W, Ma Y, Xu Y, Chang G (2020) Intermolecular channel expansion induced by cation-π interactions to enhance lithium storage in a crosslinked π-conjugated organic anode. J Power Sources 449:227551

47. Chang G, Yang L, Liu S, Lin R, You J (2015) Rational design of a fluorescent poly(N-aryleneindole ether sulfone) switch by cation-π interactions. Polymer Chem 6:697–702
48. Liu Y, Miao K, Dunham NP, Liu H, Fares M, Boal AK et al (2017) The cation-π interaction enables a halo-tag fluorogenic probe for fast no-wash live cell imaging and gel-free protein quantification. Biochem 56(11):1585–1595
49. Chen W, Elfeky SA, Nonne Y, Male L, Ahmed K, Amiable C et al (2011) A pyridinium cation-π interaction sensor for the fluorescent detection of alkyl halides. Chem Commun 47:253–255
50. Leduskrasts K, Kinens A, Suna E (2019) Cation-π interactions secure aggregation induced emission of planar organic luminophores. Chem Commun 55:12663–12666
51. Hiratani T, Konishi K (2004) Surface-cap-mediated host-guest chemistry of semiconductor CdS: intercalative cation accumulation around a phenyl-capped CdS cluster and its notable effects on the cluster photoluminescence. Angew Chem Int Ed 43:5943–5946
52. Zhong J, Li Z, Guan W, Lu C (2017) Cation-π interaction triggered-fluorescence of clay fillers in polymer composites for quantification of three-dimensional macrodispersion. Anal Chem 89:12472–12479
53. Eckel R, Ros R, Decker B, Mattay J, Anselmetti D (2005) Supramolecular chemistry at the single-molecule level. Angew Chem Int Ed 44:484–488
54. Kasprzak A, Sakurai H (2019) Site-selective cation-π interaction as a way of selective recognition of the caesium cation using sumanene-functionalized ferrocenes. Dalton Trans 48:17147–17152
55. Spisak SN, Wei Z, Rogachev AY, Amaya T, Hirao T, Petrukhina MA (2017) Double concave cesium encapsulation by two charged sumanenyl bowls. Angew Chem Int Ed 56:2582–2587